THE CITY'S MANY FACES

**Proceedings of the RAND Arroyo-MCWL-J8 UWG
Urban Operations Conference
April 13-14, 1999**

Edited by

Russell W. Glenn

with

**Geri Cherry
Lois M. Davis
Sean J. A. Edwards
Ernst Isensee
Scott Gerwehr
John Matsumura
Randall Steeb
Jamison Jo Medby**

WITHDRAWN

Prepared for the United States Army

ARROYO CENTER

RAND

PREFACE

On April 13–14, 1999, the RAND Arroyo Center, Marine Corps Warfighting Lab (MCWL), and J8 Urban Working Group (UWG) co-hosted a conference on military urban operations in the Joint Military Intelligence College auditorium, Bolling Air Force Base, Washington, D.C. The conference sought to provide a forum for information exchange and debate on the complete range of possible operations in urban areas likely to challenge U.S. national interests in the next generation. Such activities could include:

- Operations at the strategic, operational, and tactical levels of war

- Domestic and international events

- Combat, stability, and support missions

- Service, joint, multinational, and interagency concerns

- Homeland defense and WMD issues

Approximately two hundred individuals attended the conference. They included representatives from the U.S. academic, media, commercial, and public service communities, all four armed services, and several joint agencies. Military personnel from Japan, New Zealand, Norway, Spain, and the United Kingdom were also in attendance. The agenda with a list of all speakers and panel members appears in Appendix A.

Research conducted in conjunction with this document was carried out in the RAND Arroyo Center, a federally funded research and development center sponsored by the United States Army. This work was conducted in the Arroyo Center's Force Development and Technology Program for the Assistant Secretary of the Army (Acquisition, Logistics, and Technology).

This summary of proceedings will be of interest to government and commercial sector personnel whose responsibilities include policy design, funding, planning, preparation, or the development of technologies for, or the conduct of operations in, urban environments.

CONTENTS

This document presents the proceedings of an urban operations conference held on April 13–14, 1999. The event sought to educate those interested in this complex topic while simultaneously providing a forum for debate on relevant issues. The attendee who was fortunate enough to hear all twenty-eight speakers and panel members did, in fact, learn much about the many faces a city can present during military urban operations. Nearly a third of those at the podium had "seen the urban elephant" from one or more of its multitudinous perspectives.[1] Others had been students and analysts of urban operations for years. Appendix A is a copy of the agenda for the event. Subsequent appendices contain speakers' and panel members' slides and supporting notes.

After a welcome by **David Chu**, Director of the RAND Arroyo Center, **Dr. A. Michael Andrews, II** of ASA(ALT) provided a comprehensive overview of his organization's efforts to support improved soldier urban operations readiness, both in the immediate future and well into the next century. **MG Norton A. Schwartz** followed with an innovative and somewhat contentious approach to military operations in cities, one that was undeniably joint, but one that also forwarded a more wide-ranging role for the U.S. Air Force than some might envi-

---

[1]In the sense applied here, the term "see the elephant" refers to having experienced combat, especially for the first time (though the author has expanded the scope of the term to include experience with military urban operations that did not include combat in this case). Its first known use dates from the mid-19th century; the term attained common usage during the American Civil War. *Random House Historical Dictionary of American Slang, Volume 1, A–G*, 1st ed., s.v. "see the elephant."

sion. Retired **MG James D. Delk**, who served as field commander during the 1992 Los Angeles riots, identified a myriad of lessons that could benefit active and reserve military forces as well as civilian agencies in future domestic crises, be they riots, natural disasters, or other events necessitating the joining of those groups. **MG Robert Scales** challenged the audience to consider dramatically different alternatives to current urban operational doctrinal approaches. **Lieutenant Sid Heal** of the Los Angeles County Sheriff's Department addressed the topic the previous year's conference attendees had identified as the one they would have most liked to discuss more fully: the application of nonlethal capabilities during urban contingencies. The Assistant Commandant of the Marine Corps, **Gen. Terrance Dake**, provided the conference's keynote presentation. He put forth operational approaches to urban missions that dramatically differed with those described by speakers earlier in the day.

RAND's **Dr. Randy Steeb** opened the second day with a look into future urban challenges and potential changes to doctrine and technology that the U.S. armed forces might employ in confronting those demands. **LtCol Duane Schattle** of the J8 Urban Working Group provided a summary of the organization's work completed thus far and that remaining to be accomplished. A presentation of the Urban Warrior experiment's objectives and the goals of Marine Corps Warfighting Lab in general was given by **COL Gary Anderson**. He was followed by **COL Lester Martinez-Lopez**, whose personal experiences in Haiti provided the basis for a detailed and extensive identification of lessons pertinent to military medical support during stability missions in urban areas and the surrounding countryside.

These presentations by individual speakers were complemented by panel discussions in four primary subject areas. General Delk was joined by **BG Edward T. Buckley**, the FBI's **Jim Rice**, and the U.S. Army Chemical School's **COL Daniel Uyesugi** on a domestic MOUT/WMD panel. General Buckley commanded active component Army forces during the 1992 Los Angeles riots. Mr. Rice covered the wide range of considerations and organizations involved in any domestic urban operations contingency while also identifying the role the Federal Bureau of Investigation can play in overseas contingencies. Colonel Uyesugi analyzed the far-ranging implications a weapons of mass destruction (WMD) event would entail for military and civilian agencies alike. The second group, the MOUT doctrine panel, entailed six discussions. The Marine Corps' **MAJ Mark Sumner** and Army's **LTC Mark Reardon** capsulated their services' recent initiatives in urban operations doctrine. **MAJ Jeff Newell** discussed development of the pending *Handbook for Joint Urban Operations.* Two aviators, the Marine Corps' **MAJ Floyd Usry** and USAF's **MAJ Brooks Wright,** presented their services' recent advances in MOUT aviation testing, training, and experimentation. Major Sumner completed the doctrine discussion with a review of preliminary work on JP 3-06, *Joint Urban Operations.* An initiatives/technology panel provided the forum for an update on the status of the Army (**MAJ Lee Offen**) and Marine Corps (**LtCol John Allison**) Advanced Concepts Technology Demonstrations. **LTC Bob Hahn** likewise reviewed Army After Next thinking on the challenges of urban operations in the 2025 period. Two doctoral candidates representing the RAND Arroyo Center, **Sean Edwards** and **Scott Gerwehr,** respectively described their innovative work on urban communications and deception operations. The final offering of the conference was the panel on recent MOUT operation. The audience was fortunate to have an opportunity to draw on the experiences or in-depth analyses of four field experts. **COL Dave Patton's** extensive time in Haiti served as the foundation for his comments on stability and support mission tasks in urban environments. **COL T. W. Parker** similarly looked back on his noncombatant evacuation operations (NEO) responsibilities in Africa to provide lessons of value in preparing for similar and other future contingencies in cities. Two gentlemen with extraordinary insights into the Russian experience in Grozny, **Tim Thomas** and **Les Grau,** respectively analyzed asymmetry in urban warfare and logistical support during such events.

These many individuals covered much of the military urban operations spectrum, from support missions in Haiti to combat in Grozny to consideration of WMD concerns, from domestic and homeland defense matters to events worldwide, from the lowest tactical levels to matters of operational and strategic importance. In so doing they not only educated with regard to their individual specific topics, but in the larger sense they communicated that urban operations encompass far more in the way of scope and complexity than has at times been understood. Individual lessons found context in the greater whole.

ACKNOWLEDGMENTS

A special thanks is in order for RAND's co-hosts of the 1999 conference: the men and women of the United State Marine Corps Warfighting Lab and those of the Joint Staff's J8 Urban Working Group. Notable in their support were Lieutenant Colonel John Allison, Lieutenant Colonel Larry Corbett, and Corporal Cullen Waters of the Warfighting Lab, and Lieutenant Colonel Duane Schattle and Mr. Steve Marsh, J8 Urban Working Group.

For the second consecutive year, the success of the conference was very much due to the extraordinary efforts of Donna Betancourt, Director of Operations, Army Research Division, RAND Arroyo Center. Others who helped with major and minor elements of coordination, and were therefore invaluable to the event's success, include Sandy Dougharty, Phyllis Switzer, and Susan Tribotti. As always, notable thanks are due to the woman who responded to all urgent requests and ensured critical tasks were never allowed to escape completion: my secretary, Mary Aguilar.

The entire substance of the conference was attributable to the excellent presentations of the gentlemen whose names appear in the first appendix. Without these speakers' time, effort, and willingness to take on the tough problems that continue to confront our armed services in the urban operations arena, there could have been no forum for this valuable joining of men and women interested in assisting the nation as it seeks solutions. MAJ Bob Ellithorpe is deserving of special thanks for his coordinating the Assistant Commandant of the Marine Corps' presentation. To LTG Patrick Hughes, Gene Baumgartner, and all others with the Defense Intelligence

Agency who assisted in providing facilities and support at Bolling Air Force Base, RAND, the Marine Corps Warfighting Lab, and the J8 Urban Working Group express their appreciation.

The timely fashion in which this document was released and the fine quality of its presentation are entirely attributable to four highly skilled and extraordinarily dedicated women, the most able Nikki Shacklett, the superbly talented Pamela Thompson, the supremely patient Ann Deville, and the indefatigable Patrice Lester. The fine cover is once again in large part the product of work by artists Ronald Miller and Peter Soriano. Rod Sato's artwork in support of the conference proper lent much to its professional presentation. A thanks to Paul Murphy for overseeing the entire art and editorial process.

The RAND Arroyo Center's urban operations studies are very much a team effort. The editor thanks each team member for his or her help during the preparation and conduct of the affair: MAJ Geri Cherry, Sean Edwards, Scott Gerwehr, LTC Ernst Isensee, John Matsumura, Dr. Randall Steeb, and Jamie Studebaker. Others who assisted at key junctures were Dr. Lois Davis and LtCol Ken Gardiner. The team thanks Dr. David Chu for his graciously opening the event and assuming the role of primary host.

The editor also recognizes the professionalism and dedication of this document's reviewers. Robert Howe of RAND and LTG Ron Christmas (USMC, ret.) both provided valuable observations and comments of service to our readers.

ACRONYMS

1st LT	First Lieutenant
2nd LT	Second Lieutenant
3D (or 3-D)	Three-dimensional
40ID	40th Infantry Division
A05	An operational level of readiness that affects equipment worn (e.g., flak vests) and weapons status (e.g., whether rounds are chambered in a rifle)
AAA	Anti-aircraft Artillery
AAN	Army After Next
AAV	Assault Amphibian Vehicle
ACC	Air Combat Command
ACE	Advanced Concept Excursion (also Aviation Combat Element)
ACOM	Atlantic Command
ACTD	Advanced Concepts Technology Demonstration
ACV	Advanced Combat Vehicle
ADT&E	Aviation Development Tactics and Equipment

AFDO	Advanced Full Dimensional Operations
AFMIC	Armed Forces Medical Intelligence Center
AFTTP	Air Force Tactics, Techniques, and Procedures
AG	Adjutant General
AGM	Air-to-ground missile
AGR	Active Guard/Reserve
AH	Attack Helicopter
AHM	Anti-helicopter Mine
ALSA	Air-Land-Sea Application Center
AO	Area of Operations
AP	Aerospace Power
APC	Armored Personnel Carrier
ARFOR	Army Forces
ASA(ALT)	Assistant Secretary of the Army (Acquisition, Logistics, and Technology)
ASARC	Army Systems Acquisition Review Council
ASC	Air Systems Command
ASD	Assistant Secretary of Defense
ASSTC	Advanced Surgical Suite for Trauma Casualties
ATA	Airborne Target Acquisition
AWACS	Airborne Warning and Control System
AWC	Army War College
AWE	Advanced Warfighting Experiment
AZ	Arizona

B	Billion
B1	An Air Force bomber
B2	An Air Force stealth bomber
BCT	Battalion Combat Team
BDA	Battle Damage Assessment
BF	Battle Force
BG	Brigadier General
BLT	Battalion Landing Team
BSTF	Battle Shaping Task Force
BTR-80	A model of armored personnel carrier (of former Warsaw Pact manufacture)
BX	Base Exchange, a military equivalent of a department store (USN, USMC, USAF)
C/B-RRT	Rapid Response Team
C2	Command and Control
C2W	Command and Control Warfare
C3	Command, Control, and Communications
C4	Command, Control, Communications, and Computers
C4I	Command, Control, Communications, Computers, and Intelligence
C4ISR	Command, Control, Communications, Computers, Intelligence, Surveillance, and Reconnaissance
CA	Civil Affairs
CAARNG	California Army National Guard

CAS	Close Air Support
CBRIF	Chemical Biological Incident Response Force (USMC organization)
CBR-N	Chemical, Biological, Radiological-Nuclear
CBT-Q	Combat Cueing
CD	Concept Demonstration
CD	Culminating Demonstration
CDC	Center for Disease Control
CGSC	Command and General Staff College
CHEM/BIO	Chemical/Biological
CHP	California Highway Patrol (also appears as CHIP)
CIA	Central Intelligence Agency
CID	Combat Identification
CINC	Commander-in-Chief
CIT	Counterintelligence Team
CM	Consequence Management
CMC	Commandant of the Marine Corps
CMO	Civil Military Operations
CMOC	Civil Military Operations Cell
CNG	California National Guard
CNN	Cable News Network
CO	Commanding Officer
COL	Colonel
CONOPS	Contingency Operations

CONUS	Continental United States
COTS	Commercial off-the-shelf
CP	Command Post
CPR	Chairman's Program Review or Chairman's Program Recommendations
CROP	Common Relevant Operating Picture
CS	o-chlorobenzalmalononitrile, a riot control agent
CSA	Chief of Staff of the Army
CSAR	Center for Spatial Analysis and Remote Sensing
CSC	Combat Stress Control
CSEEA	Close Support End to End Assessment
CTA	Common Table of Allowances
CTP	Common Tactical Picture
CW	Conventional Warfare
DARPA	Defense Advanced Research Projects Agency
dB	Decibels
DBBL	Dismounted Battlespace Battle Lab
DCS	Deputy Chief of Staff
DCSDOC	Deputy Chief of Staff for Doctrine
DEA	Drug Enforcement Agency
DEPT	Department
DFI	As in DFI International: a research, analysis, and consulting firm
DIA	Defense Intelligence Agency
DIS	Distributed Interactive Simulation

DIV	Division
DLI	Defense Language Institute
DMSP	Defense Military Satellite Program
DOD or DoD	Department of Defense
DOMS	Director of Military Support
DOTES	Doctrine, Organization, Training, Equipment, Support and Facilities
DP	Decision Point
DPAG	Defense Planning Advisory Group
DPG	Defense Planning Guidance
DRB	Division Ready Brigade
DSC	Decision Support Center
DTLOMS	Doctrine, Training, Leader Development, Organizations, Materiel, and Soldiers
DXXI	Division XXI (Division 21)
E5	Enlisted pay grade 5 (sergeant in the U.S. Army)
ECOC	Experimental Combat Operations Center
ELINT	Electronic Intelligence
EMERCOM	Russian Ministry of Emergency Situations
EOC	Emergency Operations Center
ERT	Evidence Response Team
EUT	End User Terminal
EW	Electronic Warfare
EWTG	Expeditionary Warfare Training Group
FAC	Forward Air Controller

FAC(A)	Forward Air Controller (Airborne)
FAST	Fleet Anti-terrorism Security Team
FBI	Federal Bureau of Investigation
FBIHQ	FBI Headquarters
FCU	Future Combat Vehicle
FDA	Food and Drug Administration
FEMA	Federal Emergency Management Agency
FLIR	Forward Looking Infrared Radar
FLN	*Front de Libération Nationale*
FM	Field Manual
FMF	Fleet Marine Force
FMSO	Foreign Military Studies Office (U.S. Army)
FORSCOM	Forces Command
FRY	Federal Republic of Yugoslavia ?????
FUE	First Unit Equipped
FXXI	Force XXI (Force 21)
FY	Fiscal Year
G3	Operations, Plans, and Training staff section (U.S. Army and U.S. Marine Corps)
GCE	Ground Combat Element
GEN	General
GHz	Gigahertz
GI	Slang term for soldiers (from "government issue")
GIDB	Geospatial Information Database

GOSSIP	Ground Observation Special Support Intelligence Program
GOTS	Government off-the-shelf
GPS	Global Positioning System
GYSGT	Gunnery Sergeant
H	Hornet (a wide-area mine)
H/K	Model designator for the AGM-65 (Maverick missile). H is an electro-optical seeker with near-IR capability in a missile with an 84-pound shaped charge warhead (penetrator). The K model is the same seeker on a 300-pound warhead for blast capability.
HAZMAT	Hazardous Materials
H-Hour	Hour (time) an operation is to commence
HIV	Human Immune Deficiency Virus
HMRU	Hazardous Materials Response Unit
HNP	Haitian National Police
HOC	Humanitarian Operations Cell
HQ	Headquarters
HUMINT	Human Intelligence
I2	Image Intensifier
ICRC	International Committee of the Red Cross
ID	Identification
IFF	Identification, Friend or Foe
IGPS	Individual Global Positioning System
IMMACCS	Integrated Marine Multi-Agent Command and Control System

IMMIN	Image Intelligence
IMO	International Migration Organization
IN	Infantry
IN (L)	Infantry (Light)
IO	Information Operations
IPB	Intelligence Preparation of the Battlefield
IPSF	Interim Public Security Forces
IR	Infrared
IRA	Irish Republican Army
ISD	Information Strategy Division
ISR	Intelligence, Surveillance, and Reconnaissance
ITT	Integrated Idea Team
J2	The section on a joint staff responsible for intelligence operations
J2P	J2 Plans section
J8	Director for Force Structure, Resource, and Assessment, Joint Staff
J9	Joint Experimentation Directorate
JADD	Joint and Army Doctrine Directorate
JAG	Judge Advocate General
JASSM	Joint Air-to-Surface Standoff Missile
JCATS	Joint Conflict and Tactical Simulation
JCS	Joint Chiefs of Staff
JDAM	Joint Direct Attack Munition
JFC	Joint Force Commander

JMO-T	Joint Medical Operations-Telemedicine
JMOUT	Joint Military Operations on Urbanized Terrain
JOC	Joint Operations Center
JP	Joint Publication
JRB	Joint Review Board (???) Jeff
JROC	Joint Requirements Oversight Council
JRTC	Joint Readiness Training Center
JSIMS	Joint Simulation System
JSOTF	Joint Special Operations Task Force
JSR	Joint Strategy Review
JSTARS	Joint Surveillance, Target Attack Radar Systems
JT&E	Joint Test and Evaluation
JTF	Joint Task Force
JTF-LA	Joint Task Force-Los Angeles
JTTF	Joint Terrorism Task Forces
JTTP	Joint Tactics, Technique, and Procedures
JVAC	Joint Vaccine
JWARS	Joint Warfare System
JWCA	Joint Warfighting Capabilities Assessment
JWFC	Joint Warfighting Center
KEO	Kinetic Energy Ordnance
LA	Los Angeles
LADAR	Laser RADAR
LAN	Local Area Network

LAPD	Los Angeles Police Department
LAR	Light Armored Reconnaissance
LASO	Los Angeles Sheriff's Office
LAV	Light Amphibious Vehicle
LCPK	Low Cost Precision Kill
LEP	Laser Eye Protection
LFA	Local Federal Authority
LGB	Laser Guided Bomb
LIA	Logistics Impact Analysis
LLW	Land and Littoral Warfare
LLWAD	Land and Littoral Warfare Assessment Division
LMBF	Light Motorized Battle Force
LOC	Line of Communications
LOE	Limited Objective Experiment
Log	Logistics
LOS	Line-of-Sight
LPD	Low Probability of Detection
LPE	Low Probability of Exploitation
LPI	Low Probability of Interception
LSD	Landing Ship Dock
LST	Laser Spot Tracker
LT	Lieutenant
LTA	Limited Technical Assessment
LTC	Lieutenant Colonel

LtCol	Lieutenant Colonel (USMC and British Army form)
LTG	Lieutenant General
LtWt	Lightweight
M	Million
M&S	Modeling and Simulation
MAGTF	Marine Air-Ground Task Force
MAJ	Major
MARFOR	Marine Forces
MASCAL	Mass Casualty Event
MAWTS	Marine Aviation Weapons and Tactics Squadron
MCAS	Marine Close Air Support
MCCDC	Marine Corps Combat Doctrine Command
MCDP	Marine Corps Doctrinal Publication
MCDW	Minimum Collateral Damage Weapon
MCIA	Marine Corps Intelligence Activity
MCRP	Marine Corps Reference Publication
MCU	Marine Corps University
MCWL	Marine Corps Warfighting Lab
MCWP	Marine Corps Warfighting Publication
MDS	Mission Design Series
MECC	Mobile Expandable Container Configuration
MEDEVAC	Medical Evacuation
MEF	Marine Expeditionary Force

MEMS	Micro-Electro Mechanical Systems
METL	Mission Essential Task List
MEU	Marine Expeditionary Unit
MEU(SOC)	Marine Expeditionary Unit (Special Operations Capable)
MFC	Military Field Commander
MG	Major General
MHz	Megahertz
MI	Military Intelligence
MIT	Massachusetts Institute of Technology
MMIM	Micro-controllable Multifunctional Interface Module
MOE	Measure of Effectiveness
MOH	Ministry of Health
MOOTW	Military Operations Other Than War
MOPMS	Modular Pack Mine System
MOUT	Military Operations on Urbanized Terrain
MP	Military Police
MRE	Medical Rules of Engagement
MSGT	Master Sergeant
MSN	Mission
MTLB	Multipurpose Armored Vehicle of former Warsaw Pact manufacture
MTT	Mobile Training Team
MTW	Major Theater of War

MUP	Yugoslav Interior Ministry of Interior Police
MV-22	Marine version of the Osprey tilt-rotor aircraft
MVD	Russian Ministry of Internal Defense
MWSS	Marine Wing Support Squadron
NATO	North Atlantic Treaty Organization
NAVAIRPAC	Naval Air Pacific
NBC	Nuclear, Biological, Chemical
NCA	National Command Authority
NCC	Noncombatant Control
NCO	Noncommissioned Officer
NCOIC	Noncommissioned Officer in Charge
NDBS	No Drop Bomb Set System
NDIA	National Defense Industrial Association
NDP	National Defense Panel
NDPO	National Domestic Preparedness Office
NDU	National Defense University
NEO	Noncombatant Evacuation Operation
NFO	Naval Flight Officer
NGIC	National Ground Intelligence Center
NGO	Nongovernmental Organization
NIMA	National Imagery and Mapping Administration
NIPC	National Infrastructure Protection Center
NLOS	Non-line-of-sight
NLW	Nonlethal Weapons

NOAA	National Oceanic and Atmospheric Administration
Non-LOS	Non-line-of-sight
NPGS	Naval Postgraduate School
NRO	National Reconnaissance Office
NSFS	Naval Surface Fire Support
NSSG	National Security Study Group
NVG	Night Vision Goggles
NW	Nodal Warfare
NYC	New York City
OCSW	Objective Crew-Served Weapon
OES	Office of Emergency Services
OH	Operational Handbook
OICW	Objective Individual Combat Weapon
OKH	*Oberkommando des Heeres* (German Army headquarters during World War II)
OLS	Operational Linescan System
OOTW	Operations Other Than War
OPLAN	Operations Plan
OPNET	A computer simulation used for communications analysis
OPORD	Operations Order
Ops	Operations
OPTEC	Operational Test and Evaluation Command
ORC	Operational Research Consultants

OSCE	Organization for Security and Co-operation in Europe
OSD	Office of the Secretary of Defense
PAHO	Pan American Health Organization
PCS	Personal Communications Services
PDD	Presidential Decision
PDF	Panamanian Defense Force
PFC	Private First Class
PGM	Precision Guided Munition
PGMM	Precision Guided Mortar Munition
PIRA	Provisional Irish Republican Army
PLGR	Precision Lightweight GPS Receiver
PMO	Provost Marshal Office
POA&M	Programs, Objectives, Actions and Milestones
POAM	Plan of Action Milestones
POL	Petroleum, Oils, and Lubricants
POM	Program Objective Memorandum
PROWPU	Portable Reverse Osmosis Water Purification System
Pub	Publication
PVO	Private Volunteer Organization
PX	Post Exchange, a military equivalent of a department store (USA)
Q	Quarter
QDR	Quadrennial Review

R&D	Research and Development
RAID	Rapid Assessment Initial Detection
RAP	Rocket-assisted Projectile
RDs	Rounds
RDW	Radiation Dispersal Weapon
Ret	Retired
RF	Radio Frequency
RH	Relative Humidity
RLEM	Rifle-Launched Entry Munition
RMA	Revolution in Military Affairs
RML	Revolution in Military Logistics
ROE	Rules of Engagement
ROTS	Ruggedized off-the-shelf
RPG	Rocket Propelled Grenade (also used to refer to the system for firing such a grenade)
RSTA	Reconnaissance, Surveillance, Targeting, and Acquisition
RTF	Regional Task Force
RTIDS	Rapid Deployment Intrusion Detection System
RUC	Royal Ulster Constabulary
RW	Rotary Wing
S&T	Science and Technology
SAC	Study Advisory Committee
SAD	State Active Duty
SAM	Surface-to-Air Missile

SAR	Search and Rescue
SARDA	Assistant Secretary of the Army (Research, Development and Acquisition)
SARGE	Surveillance and Reconnaissance Ground Equipment
SAS	Situational Awareness System
SATCOM	Satellite Communications
SCBA	Self-Contained Breathing Apparatus
SCRAG	Senior Civilian Representative of the Attorney General
SEAD	Suppression of Enemy Air Defenses
SEV-B2	Surface Effects Vehicle, Model B2
SF	San Francisco
SGT	Sergeant
SIGINT	Signals Intelligence
SINCGARS	Single Channel Ground to Air Radio System
SIOC	Strategic Information Operations Center
SO/LIC	Special Operations/Low Intensity Conflict
SOCOM	Special Operations Command
SOF	Special Operations Forces
SOI	School of Infantry
SOP	Standing Operating Procedures
SOTG	Special Operations Training Group
SP	Self-propelled
SPMAGTF(X)	Special Purpose MAGTF (Experimental)

SQ	Square
SS	Sector and Seal
STX	Situational Training Exercise
SUO	Small Unit Operations
SWAT	Special Weapons and Tactics
T&E	Training and Equipment
T-80	A tank model manufactured by nations of the former Warsaw Pact
TAC	Terminal Access Controller (also in common use: Theater Air Controller)
TACP	Tactical Air Control Party
TACTS	Tactical Aircrew Training System
TDA	Table of Distribution and Allowance
TES	Test and Evaluation Squadron
TF	Task Force
TLAM	Tomahawk Land Attack Missile
TO&E (or TOE)	Table of Organization and Equipment
TOW	Tube-launched, optically-tracked, wire-command link guided missile
TPFDD	Time-Phased Force Deployment Data
TPFDL	Time-Phased Force Deployment List
TPO	Technology Program Office
TRAC	TRADOC Analysis Center
TRADOC	Training and Doctrine Command (U.S. Army)

TSG	Technology Seminar Game
TTP	Tactics, Techniques, and Procedures
UAV	Unmanned Aerial Vehicle
UCA	Urban CAS Assessment
UCAS	Urban Close Air Support
UCLA	University of California, Los Angeles
UFR	Unfunded Requirement
UGV	Unmanned Ground Vehicle
UH	Utility Helicopter
UHF	Ultra High Frequency
UN	United Nations
UNHCR	United Nations Office of the High Commissioner for Refugees
UNICEF	United Nations International Children's Emergency Fund
UNMIH	United Nations Mission in Haiti
UNOSOM	United Nations Operations, Somalia
UP	Urban Preparation
US	United States
USA	United States Army
USAF	United States Air Force
USAID	United States Agency for International Development
USAWC	United States Army War College
USC	University of Southern California

USCG	United States Coast Guard
USMC	United States Marine Corps
USN	United States Navy
USS	United States Ship
UTC	Urban Target Complex
UW	Urban Warrior or Unconventional Warfare
UWB	Ultra-Wideband
UWG	Urban Working Group
UXO	Unexploded Ordnance
V&V	Verification and Validation
VHF	Very High Frequency
VIP	Very Important Person
VIT	Virtual Interactive Targeting/Virtual Interactive Targets
VMDS	Video Motion Detection System
VTA	Russian Military Transport Aviation
WAM	Wide Area Mine
WFO	Washington Field Office
WFOV	Wide Field of View
WFP	World Food Program
WHO	World Health Organization
WMD	Weapons of Mass Destruction
WRAP	Warfighting Rapid Acquisition Program
WSO	Weapons System Officer

WTI	Weapons and Tactics Instructor
WW2	World War Two
Y2K	Year 2000, often used to refer to computer problems relating to the changing of the year from "99" to "00"
ZOS	Zone of Separation

Chapter One

INTRODUCTION

Heterogeneous *adj* [ML *heterogeneus,* fr. Gk *heterogenes,* fr. *Heter-* + *genos* kind – more at KIN] (1630): consisting of dissimilar or diverse ingredients or constituents: MIXED

—*Mirriam-Webster's Collegiate Dictionary,* 10th ed.

There is perhaps no single word that better describes a modern city than the one above. The topography is layers-on-layers of manmade and once natural terrain. Folds in the ground and pre-urban water-ways appear and disappear among structures of virtually every conceivable sort, some beneath the ground, most on it, many towering into the skies above. The functions served by these structures are as varied as their forms, their materials and mode of construction, and their appearance. The majority of these roles demand the attention and care of the urban area's human occupants. Other tasks are partially or fully automated, requiring only the occasional, if any, intrusion by man. The city's inhabitants are themselves no less diverse than are its physical attributes. Some are residents; many depart nightly; the presence of others is more transient yet, based on the demands of travel, leisure, or responsibility. Regardless of the time or day or season, the density of human occupants far exceeds the norm found in other locations. Were the variations and density of the humans not enough, the metropolis is likewise the habitat for myriad other animals, some domesticated, a few indigenous, and many capable of transmitting disease given the opportunity.

This heterogeneity is common to all cities; they are alike in being so internally dissimilar. The variation is but one defining feature of urban areas; the speed with which that heterogeneity changes is another. Urban areas are constantly undergoing metamorphosis. Further, the rate of transformation tends to increase in times of stress. The composition of a municipality's population can be dramatically altered by the influx or egress of refugees, by deliberate displacement, or by disease or conflict-related disasters. Alteration of the physical space can undergo similar mutation in very limited periods. Mountains and hills change over eons, fields and forests over decades; cities can transform in days. This pace of urban change joins heterogeneity and high population density to make today's cities the most difficult of military operational areas. Nations considering the commitment of men, women, and resources into one or more urban areas must understand that preparedness demands flexibility. The capability to adapt to the unexpected may be more important than any other measure of readiness.

The challenges confronting a military readying for such contingencies are themselves characterized by extreme diversity. Forces must be prepared to meld the objectives of the tactical, operational, and strategic levels of war, and to do so both internationally and domestically during combat missions and support or stability taskings that may or may not include fighting. Such forces could conceivably find themselves offering basic life-giving aid, subduing riotous crowds, clashing with enemy warriors, and attempting to contain the effects of a biological, chemical, or nuclear event within a single urban conglomeration at a given instant in time. The undertaking for any one of the U.S. armed forces would very likely be both abetted and complicated by the simultaneous necessity to work with other of the nation's services, the militaries of coalition members, and a considerable number of governmental and private agencies.

On April 13–14, 1999, RAND, the Marine Corps Warfighting Lab, and the J8 Urban Working Group joined forces to host a conference dedicated to taking a further step toward resolving these many issues. The goal was to educate and provide a forum for debate. A similar event had attracted nearly seventy attendees the year before.[1] It is indicative of the still-increasing interest in the urban operations field that the 1999 affair drew some two hundred individuals. A review of the events in the fourteen months between the two conferences similarly reflects the enhanced level of concern. In April 1998, the Marine Corps released its MCWP 3-35.3, *Military Operations on Urbanized Terrain*, replacing a 1980 document.[2] By October 1998 the Joint Doctrine Working Party had approved initiation of the U.S. armed forces' first joint urban operations publication (JP 3-06). By year's end the *Handbook for Joint Urban Operations* was being created to assist U.S. forces should urban contingencies arise prior to completion and approval of JP 3-06.

The pace at which exercises, experiments, and conferences were held also saw an increase. The series of tests, demonstrations, and experiments that comprised the Marine Corps' Urban Warrior culminated in February 1999 with an event in the Oakland–San Francisco Bay area that included Navy and Coast Guard participation in addition to that of the ground force. The Air Force, sponsoring the writing of the *Handbook for Joint Urban Operations*, held a conference in March 1999 as an initial step in charting its course toward greater urban operations preparedness in coming years. The Special Operations Command had similarly conducted its Urban Operations War Game the month before. The 1999 Army After Next Spring War Game spanned the conference and had a significant urban focus. Unlike in previous years, the high technology test force was committed to city fighting during the ultimate week of analyzing possible solutions to the challenges of the 21st century. The Army and Marines moved toward completion of their MOUT Advanced Concepts Technology Demonstrations (ACTD) and a joint exercise to put new concepts to the test in a field environment at Fort Polk's Shugart-Gordon urban training facility. Myriad additional projects, conferences, and training events featured increased emphasis on military operations in urban areas. After decades of neglect, the U.S. armed services had

---

[1]See Russell W. Glenn et al., *Denying the Widow-maker: Summary of Proceedings, RAND-DBBL Conference on Military Operations on Urbanized Terrain*. Santa Monica, CA: RAND, CF-143-A, 1998, for coverage of the 1998 conference.

[2]In January 1998, a month before the first conference, the U.S. Army decided to rewrite its primary urban operations field manual, FM 90-10, also entitled *Military Operations on Urbanized Terrain*.

begun to reacquaint themselves with what had become seemingly inevitable urban contingencies.[3]

By the time of the April 1999 conference, few were unaware of increasing worldwide urbanization. It seemed that every article or report on military operations in urban areas cited the dramatic facts: a predominantly urban world in the opening years of the 21st century, a more rapid rate of urbanization in developing vice developed nations, and the increasing influence of cities as economic, social, cultural, transportation, and political hubs. Only those unwilling to see failed to recognize that this concentration of influence, the proximity of airfields and ports to urban areas, and the United States' reliance on military force projection would inevitably compel America's military men and women to partake of urban combat. Panama City, Mogadishu, Port-au-Prince, Brcko, Monrovia, and Freetown were but a sampling of the world's metropolises that had seen U.S. forces in their streets during the decade before. Equally as obvious were shortfalls in the nation's training facilities and technological preparedness as the armed forces prepared to face the challenges of urban contingencies more demanding than these. Considerable differences notwithstanding, the nation had lessons to learn from the Russians' humiliation in Grozny in addition to its own experience in 1993 Mogadishu; the United States could not afford to be complacent.

Less immediately daunting, perhaps, but no less important, were the expanding threats to domestic towns, suburbs, and cities. Active and reserve armed forces had to be ever-ready to assist in times of natural (e.g., earthquake, tornado, hurricane, flood) or manmade (riots, WMD, terrorist bombing) disasters. Exercises linking domestic and Department of Defense agencies with the objective of building understanding and consensus were rare. Operations during the 1992 Los Angeles riots and New Madrid fault earthquake exercises at the Army's School of Advanced Military Studies demonstrated that much coordination was necessary. Some lessons were fundamental: military units that considered the boundaries of police and fire districts when assigning areas of responsibility found their coordination tasks

---

[3]For a listing of further armed forces initiatives in this regard, see Russell W. Glenn. "Fox Trot: Seeking Preparedness for Military Urban Operations." *Armed Forces Journal International*, May 1999, pp. 46–49.

much simplified compared to those that looked at topography alone. Others were highly complex: issues regarding who should head domestic coalitions of military, police, fire, legal, city/county/state/ federal political, and community organizations were often resolved only because of exceptional selflessness and dedication to task accomplishment. It is naïve to rely on such sacrifices during future contingencies.

The objectives of the RAND-MCWL-J8 UWG conference were to educate and stimulate debate. If it accomplishes the additional feat of having spurred cooperation and greater attention to the topic, it has fortuitously benefited the nation's soldiers, sailors, marines, airmen, and other citizens all the more.

JOINT URBAN OPERATIONS
C O N F E R E N C E

RAND Arroyo Center, Marine Corps Warfighting Lab, J8 Urban Working Group

THE CITY'S MANY FACES: INVESTIGATING THE MULTIFOLD
CHALLENGES OF URBAN OPERATIONS

Joint Military Intelligence College Auditorium (Building 6000)
Bolling Air Force Base, Washington, D.C.

13–14 April 1999

Tuesday, 13 April

0730–0800   **Registration** (DIAC Lobby)

Continental Breakfast (Bowman Room)
(Coffee and danish available throughout the morning)

0800–0815   **Welcome and Introduction**
Dr. David S. C. Chu, Vice President, Army Research
Division and Director, RAND Arroyo Center

0815–0900   **The Army S&T Program and its Urban Operations**
Dr. Michael Andrews, Deputy Assistant Secretary
of the Army for Research and Technology

0900–1000   **The Role of Aerospace Power in Joint Urban
Operations**
Major General Norton A. Schwartz
Director of Strategic Planning, Deputy Chief of Staff
for Plans and Programs, U.S. Air Force

1000–1015   **Break**

1015–1115   **MOUT: A Domestic Case Study—The 1992 L.A. Riots**
Major General James Delk (CAARNG, Ret.)

1115–1230   **Domestic MOUT/WMD Panel**

The CAARNG and the 1992 Los Angeles Riots
Major General James Delk (CAARNG, Ret.)

**The Active Army Component and the 1992 Los
Angeles Riots**
Brigadier General Edward Buckley, Deputy Chief of
Staff for Doctrine, U.S. Army Training and Doctrine
Command

**FBI Concerns During Domestic Military Urban
Operations**
Mr. James W. Rice, Supervisor, Special Agent,
Federal Bureau of Investigation

WMD in Urban Operations
Colonel Daniel F. Uyesugi, Assistant Commandant,
U.S. Army Chemical School

1230–1330   **Lunch** (Cafeteria)

Tuesday, 13 April (continued)

1330–1445   **MOUT Doctrine Panel**

U.S. Marine Corps MOUT Doctrine (MCWP 3-34.3)
Major Mark D. Sumner, Infantry Officer, Ground
Doctrine Branch, Doctrine Division, Marine Corps
Combat Development Command

U.S. Army MOUT Doctrine (FM 90-10)
Lieutenant Colonel Mark Reardon, Chief Special
Doctrine Team, U.S. Army Command and General
Staff College

U.S. Air Force MOUT Doctrine (Joint MOUT Handbook)
Major Jeff Newell, Staff Officer, Strategic Planning
Directorate, U.S. Air Force

U.S. Marine Corps Aviation Doctrine
Major Floyd Usry, AH-1W Instructor, U.S. Marine Corps

U.S. Air Force MOUT Testing
Major Brooks Wright, Chief, A-10 Test Division, 422
Test and Evaluation Squadron, U.S. Air Force

Joint MOUT Doctrine (JP 3-06)
Major Mark Summer

1445–1500   **Break**

1500–1600   **A MOUT Doctrinal Concept**
Major General Robert Scales, Jr., Commandant, U.S.
Army War College

1600–1700   **Nonlethal Options: Failings and Futures**
Lieutenant Sid Heal, Special Projects Group, Los
Angeles Sheriff's Department

1700–1715   **Concluding Remarks for Day 1**

1745–2130   **Dinner** (Ballroom, Bolling Officers' Club, 50 Thelsen St.)

1745   No-Host Bar

1830   Buffet Dinner

1915   **The City's Many Faces**
General Terrance Dake, Assistant Commandant, U.S.
Marine Corps

Wednesday, 14 April

0730–0800   **Continental Breakfast** (Bowman Room)
(Coffee and danish available throughout the morning)

0800–0845   **Preemption for MOUT: Revisiting an Old Strategy**
Dr. Randall Steeb, Senior Scientist, RAND

0845–0930   **Joint MOUT Mission Area Analysis and Mission Need Assessment**
Lieutenant Colonel Duane Schattle, Principal Advisor, J8 Urban Working Group

0930–1030   **Urban Warrior and U.S. Marine Corps Urban Operations**
Colonel Gary Anderson, Chief of Staff, Marine Corps Warfighting Lab

1030–1100   **Break**

1100–1200   **Medical Support for Urban Operations**
Colonel Lester Martinez-Lopez, Commander, Martin Army Community Hospital

1200–1300   **Lunch** (Cafeteria)

1300–1430   **Initiatives/Technology Panel**

U.S. Army ACTD
Major Lee Offen, Dismounted Battlespace Battle Lab MOUT ACTD Plans Officer

U.S. Marine Corps ACTD
Lieutenant Colonel John Allison, Director, Experimental Operations Division, Marine Corps Warfighting Lab

Communications in Urban Environments
Mr. Sean Edwards, Researcher, RAND

Urban Operations and the Army After Next
Lieutenant Colonel Robert Hahn, Deputy Chief of Staff-Doctrine, U.S. Army Training and Doctrine Command

Deception in the City
Mr. Scott Gerwehr, Researcher, RAND

Wednesday, 14 April (continued)

1430–1445   **Break** (Cookies, coffee, and soft drinks available in the Bowman Room)

1445–1600   **Recent MOUT Operations Panel**

Haiti
Colonel Dave Patton, Commander, Fort Myer, U.S. Army

NEOs
Colonel Thomas W. Parker, HQDS, Manpower and Reserve Affairs, U.S. Marine Corps

Grozny I: Asymmetry in Urban Warfare
Mr. Timothy Thomas, Analyst, Foreign Military Studies Office, U.S. Army

Grozny II: Logistical Support During Urban Operations
Mr. Lester Grau, Military Analyst, Foreign Military Studies Office, U.S. Army

1600–1615   **Concluding Remarks**

THE ARMY S&T PROGRAM AND ITS URBAN OPERATIONS INITIATIVES
Dr. A. Michael Andrews, SARDA

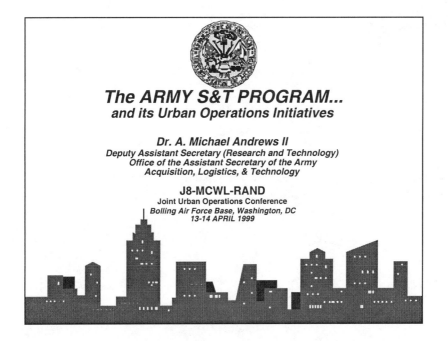

Complex terrain is a real problem that both the Marines and Army face. I want to tell you a little bit about the Army's Science and Technology Program and investment strategy before I launch into its urban operations initiatives. Over the next five years, Science and Technology investment in support of Army dismounted soldier efforts will be on the order of approximately three-quarters of a billion dollars. The question is, of course, in a time of limited resources, do we need to do more or should we do less? Today I'll tell you a little bit about how we get to answering some of these questions.

Discussion

- **Army S&T Investment Strategy**

- **Army S&T Initiatives**

- **Urban Operations S&T Initiatives**

- **Bottom Line**

Marines conducting building clearing operations during
MOUT ACTD Marine Corps Experiment 3 at Camp Lejeune

Soldiers in four man stack outside building at Fort Benning's
McKenna MOUT training site during MOUT ACTD Army
Experiment 2.

I'd like to describe our investment strategy first. How do we make the right decisions for the future? The future is very hazy when you look out there in the 2025 time frame. These photos of actions at Camp Lejeune and Fort Benning show some of the activities associated with MOUT ACTD experiments. At the core, our work is supporting marines and soldiers like these. Our ten thousand Army scientists and engineers are part of that support process.

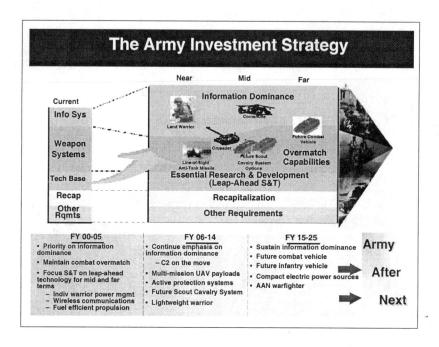

This shows you the Army investment strategy. Many of you have probably seen this. I've overlaid it with a few pictures that reach out into the Army After Next time frame. Coming out of the Cold War environment, we need to rethink our investment strategy. We have a lot of legacy systems. Apaches, for example, will look about the same in another fifteen to twenty years, but will have undergone a brain transplant. That is, in part, what information dominance is about: making sure our soldiers have more capability than what they presently possess. With information dominance we will have a better-integrated battlefield.

There are several key elements to mention in this regard. One is Land Warrior. Another piece of the solution in the mid-time frame is the Future Scout and Cavalry System. It is both an informational and overmatch system that will provide better situational awareness on the battlefield. Finally, you reach out to the far future. TRADOC is developing a mission need statement for a future combat vehicle and other capabilities, everything from nonlethals to direct and indirect fires to means of carrying troops to operating in urban terrain. The problem, of course, is that we will confront a variety of scenarios. You cannot say we're only going to need a heavy force, nor can you plan on only having requirements for a light force. We will have to have all those capabilities to some degree.

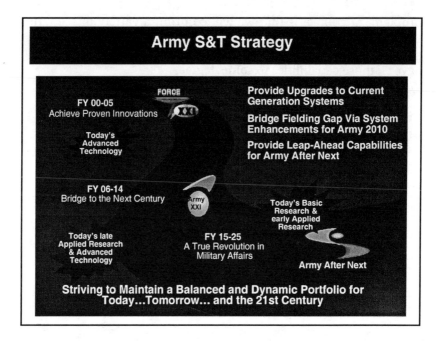

Editor's note: *The slide makes use of terminology employed in congressional budget line items. "Basic Research" involves fundamental scientific studies to examine new concepts. "Applied Research" investigates the feasibility of concepts with high military applicability. "Advanced Technology" demonstrates technical feasibility and develops technologies for transition into Program Definition and Risk Reduction and Engineering Manufacturing and Development programs.*

From the overall Army investment strategy comes the Science and Technology (S&T) strategy. Today's applied technology program is going to really pay off in the 2010 timeframe. An example of that payoff is the Future Scout and Cavalry System. It is being developed with a fast track process; the other services are watching to see how well we're able to accomplish that. We will move into First Unit Equipped (FUE) status in about five to six years. Normally the process takes twelve years to achieve fielding. At the same time, we have to be able to decide to kill projects when it appears the warfighter doesn't need them. Overall, it's about trying to reach a balance.

Changing Environment for Technology

- **High Expectations for Future Military Capabilities***
  - **3x more effective**
  - **3x more mobile**
  - **1/3 of today's support**
  - **Greater dependence on commercial technologies**

- **AAN Contingency Forces-** enabled by leap-ahead technologies
  - **Rapidly deployable**
  - **Lethal forces**
  - **Conduct combat operations for limited periods without resupply**
  - **Survivable**
  - **Versatile - able to fight/deploy in various environments**

- **Greater Reach Expected for Military Technology Investments**
  - **Accelerated cycle Times**
  - **Quantum improvements in products between cycles**
  - **Global phenomenon - worldwide marketplace**

* ASB Summer Study Briefing, October 1998

Let me talk a little bit about the changing technology environment. Last summer's Army Science Board talked about expectations for future military capabilities. Some of those have been embraced in the future combat vehicle. We are looking for future Army combat vehicles to be more effective, more mobile, and less demanding of support. (In fact, one of our goals is to approach a 90 percent reduction in sustainment costs.) We need to be more dependent on commercial technologies. These will be complemented by what are called Leap Ahead technologies, those advances you normally cannot project. These pose discontinuities. Off-the-shelf, military procurement, and Leap Ahead technologies must give us the capabilities needed to meet the requirements of tomorrow's force projection Army. If you have a 20ish-ton vehicle facing a 70-ton tank, you've got to be able to hold your own ground and win. We've got to make sure there's that kind of overmatch in our development.

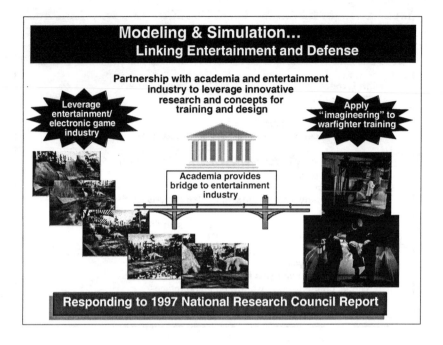

An area of promise, but one requiring considerable work, is urban simulations. We are working with operators, the academic community, and other pertinent sectors in our efforts to provide better training and analysis simulations.

One of the important ways we get to the future is through modeling simulation. There was a 1997 national research council report that studied possible cooperation between the entertainment and defense industries. Where might they overlap? The report identified several key areas where both had the same problems, but the authors couldn't figure out how the two could ultimately work together. What we are in the process of putting in place right now is a university-affiliated research center that will provide a bridge between entertainment, which means the game industry as well as theme park–based people, and the Army. We've looked at Berkeley, USC, and UCLA as the primary players and are very close to making a final decision. The purpose here is to improve our training and design capabilities. The entertainment industry, like the Army, wants their customers to really believe they're involved. We want our soldiers to really believe that they're part of the process. We don't want them looking at their watches saying "When is this training mission over?" Further, the solution has to allow thousands of players on line at one time. Our goal is to have thousands of soldiers on line involved in a tank training exercise.

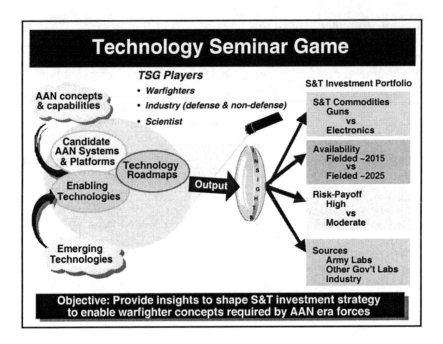

Let's go back to the question of commercial and other areas of investment. That's one reason we conduct technology seminar games. We held our first one last year. We bring in concepts from the Army After Next. We bring a unique set of players to this game: warfighters, Army and other lab scientists, and industry. If the war game is successful, we obtain a good idea of how we need to design our investment portfolio. People like the National Reconnaissance Office (NRO) and Central Intelligence Agency (CIA) are part of this investment strategy game. We also have a senior review group that examines insights for us. Last year Bill Perry was the chairman of that contingent. This year Norm Augustine has agreed to take on that responsibility, and we have some high-level players to help us maximize what we extract from game insights.

Fortunately, we don't have frequent wars to test our concepts, but this places difficult demands on the development system. You have to make your investments carefully. There's another Defense Science Board report coming out soon. I've seen the interim report, and it highlights the advantages of being able to procure parts from the commercial environment. But it also advises caution. A commercial vendor could choose to put a virus in critical software designed to activate at the initiation of hostilities. It is therefore essential that we do full system-to-system testing to ensure there are no hidden difficulties that could endanger our soldiers on future battlefields. There are thus both potential penalties and rewards in using commercial technologies.

How do we decide what is the right investment? Our journey to the future has to be able to capture what tomorrow's soldiers really need. Our challenge is to properly prepare our Army for the year 2020. It is a tough challenge. Not many people can clearly see where you need to go twenty to twenty-five years from now.

Why Invest in the MOUT Problem?

How do U.S. troops gain the technological advantage in complex terrain?

"In one moment in time, our service members will be feeding and clothing displaced refugees - providing humanitarian assistance. In the next moment, they will be holding two warring tribes apart - conducting peacekeeping operations. Finally, they will be fighting a highly lethal mid-intensity battle. All on the same day, all within three city blocks. It will be what we call the three block war."

General Charles C. Krulak
Commandant
United States Marine Corps

I show a quote from General Krulak. He talks about the three-block war. He discusses three mission-types for which our soldiers and marines must be prepared when they enter any city. We did it in Somalia; we might in Kosovo.

The MOUT Challenge

"...the most complex and resource intensive battlefield in the 21st Century..."

"...a military unprepared for urban operations across the spectrum is unprepared for tomorrow..."

"...the most likely battlefield in the 21st Century..."

Black Hawk Down

It's The Environment...!
- Noncombatants
- Electro-Magnetic Interference
- Densely Compacted Areas
- High Rise Buildings
- Modern Construction

Mark Bowden's "Black Hawk Down: A Story of Modern War"

"For sheer gore and hellishness, nothing approaches the agonies of the ground convoy as it gets lost in the maze of city streets and is cut to pieces by enemy fire– all with aircraft overhead offering useless directions by radio."

NY Times Book Review, 14 Mar 99

Why invest in the MOUT problem, one might ask? Are today's soldiers prepared for that challenge? Are they lethal enough? Are they sufficiently survivable? Can they carry out the necessary variety of missions?

For those of you who have not read *Black Hawk Down*, I recommend it. We lost eighteen soldiers killed in the fighting of October 3–4, 1993. What was planned as a two-hour mission lasted the remainder of the day and all that night. The MOUT challenge is real. It's a tough one, and it's all about the environment that soldiers and marines may have to go into.

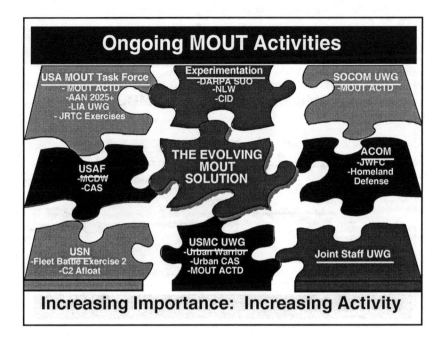

Let me talk about the overall increase in MOUT activity. This is the standard puzzle chart. How do all the pieces fit? There is a lot going on. The Army, Marines, Navy, and Air Force are involved. The Joint Staff and ACOM are attempting to assist in achieving solutions to the challenges of MOUT. Part of what is needed, and what this conference is all about, is synchronization and collaboration in efforts to find answers. If we don't work together, then we miss opportunities to maximize the benefits of the investments we're making today. It's important for us to reach out from the S&T and developer communities to make sure we know where to put our dollars to seize available opportunities.

MOUT ACTD Mission

Improve the operational capabilities of Soldiers and Marines in MOUT.

- *Evaluate advanced technologies to provide technological dominance in MOUT.*

- *Provide interim capabilities to operational units with TTPs.*

- *Set the stage for rapid acquisition of selected technologies.*

Soldiers maneuver through Fort Benning's McKenna MOUT training site during MOUT ACTD Army Experiment 2.

These opportunities include the Advanced Concept Technology Demonstration program, which includes over $60 million dollars in funding. It began in late 1997 and will wind up in the 2001–2002 time period. The objective of this program is improvement of MOUT operational capabilities for the nation's soldiers and marines. Another important element is the program's role as a vehicle for rapid acquisition of selected technologies.

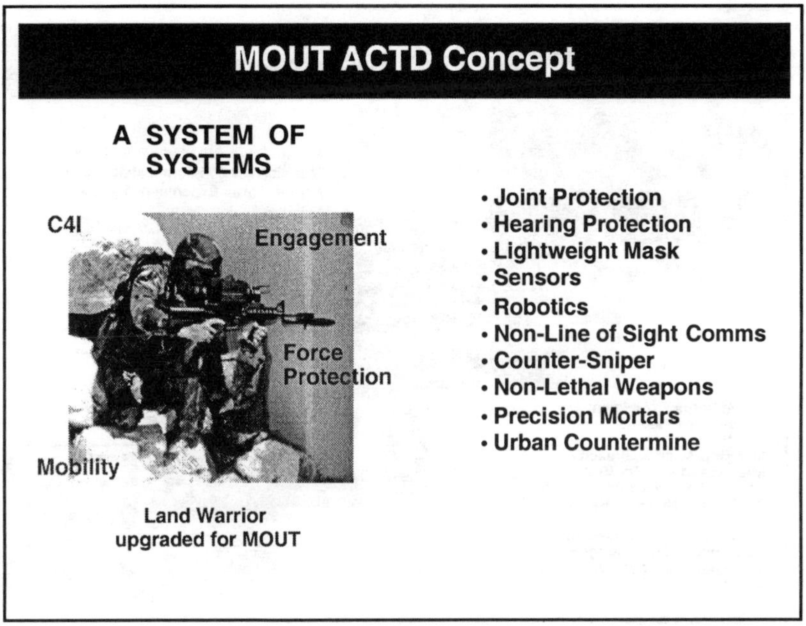

MOUT ACTD Concept

A SYSTEM OF SYSTEMS

C4I

Engagement

Force Protection

Mobility

Land Warrior upgraded for MOUT

- Joint Protection
- Hearing Protection
- Lightweight Mask
- Sensors
- Robotics
- Non-Line of Sight Comms
- Counter-Sniper
- Non-Lethal Weapons
- Precision Mortars
- Urban Countermine

Most of the Army's ACTD initiatives are really systems-of-systems. The Rapid Force Projection Initiative was also very much a systems-of-systems effort with a half-dozen major platforms and technologies involved.

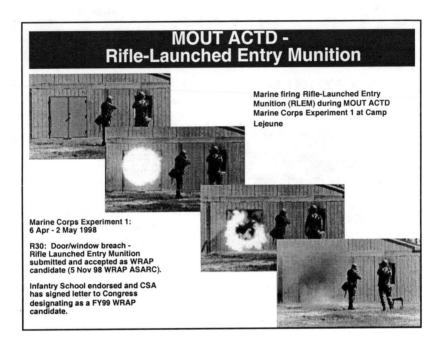

**MOUT ACTD -
Rifle-Launched Entry Munition**

Marine firing Rifle-Launched Entry Munition (RLEM) during MOUT ACTD Marine Corps Experiment 1 at Camp Lejeune

Marine Corps Experiment 1:
6 Apr - 2 May 1998

R30: Door/window breach - Rifle Launched Entry Munition submitted and accepted as WRAP candidate (5 Nov 98 WRAP ASARC).

Infantry School endorsed and CSA has signed letter to Congress designating as a FY99 WRAP candidate.

I'll give you one specific example. This is an ongoing ACTD. One of the requirements in this program is stand-off breaching of a door. The Rifle-Launched Entry Munition was so successful that we moved it into what we call a Warfighter Rapid Acquisition Program to hasten its availability to our soldiers and marines in the field

---

Advanced Concept Excursion (ACE)
MOUT ACTD

ACE .... an event to demonstrate emerging technologies not mature enough to be in the mainstream MOUT ACTD experimentation

- **augment the COTS/GOTS and more mature technology focus of the MOUT ACTD.**
- **begin formulation of an S&T investment strategy to support future Army and Marine Corps doctrinal/ operational thrusts.**
- **influence the FY02 POM S&T investment strategy for urban operations**

Shaping the Budget Through Experimentation

Most of the technologies incorporated into the MOUT ACTD are fairly mature. What we're not getting at, but would like to, are some of the more advanced technologies. We want to augment the activities that are under way right now to develop S&T technology strategies to look beyond the immediate ACTD period and into the years immediately following.

AAN Warfighter

POTENTIAL SYSTEM CHARACTERISTICS:

- total situational awareness
- no information overload
- an array of weapons (lethal and nonlethal)
- tiny, lightweight, powerful, voice-activated computer
- robotic systems
- light, nonbulky "smart" clothing
- embedded personal status monitor
- extremely mobile
- *light* load
- new power sources
- nutrient dermal sustainment patches
- individual water purification kit

Let me now shift to the more distant future and how we might get there. This concept is based on an article that will soon come out in *Parameters* [summer 1999 issue]. The idea is the future warrior. Some elements are not that much different than those forwarded in Robert A. Heinlein's *Starship Troopers*. The challenge for the S&T community is to take ideas and turn them into future capabilities. But we've got to be careful. The concepts have to be feasible. They have to be affordable. If I can only buy one of these soldier outfits, it sure won't do the rest of us any good. It has to be affordable while not violating the laws of physics. At the same time, we have to be achieving discontinuity thinking.

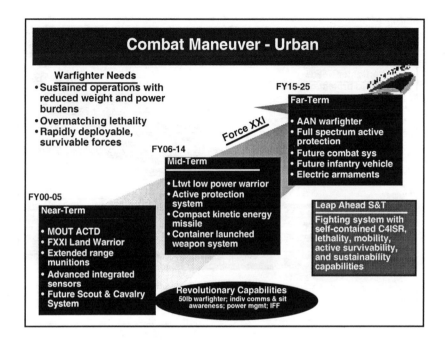

These charts address how we build our MOUT investment strategy. Right now we seek a future combat system that can meet all of TRADOC's urban battlefield requirements. The Land Warrior system is to be on the order of maybe seventy-five pounds. Given that we can achieve that, we need to look to the future and balance soldier needs with feasible advances. Perhaps we can attain a fifty-pound system capable of seventy-two hours of independent operation in a mission environment.

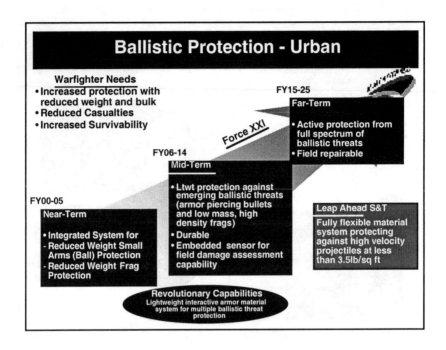

One key requirement is survivability. We've made some significant investments in S&T to meet survivability needs. An initial effort involves a system with protection of all but the neck and underarm areas. It weighs about twenty-five pounds. However, that is the protection system alone. Can we get that down a lot further while providing yet more protection? You can see that we are already looking well beyond Land Warrior. Our objective is to get our S&T applied research investment programs in shape so that we can launch another advanced technology program oriented on the 2004–2005 time frame, perhaps a little later if necessary.

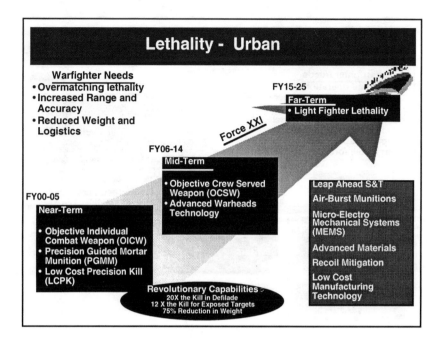

Lethality in the urban environment is another area of concern. The Objective Individual Combat Weapon (OICW) will graduate from S&T in 1999 and move into its development program. The OICW is well along that route, but it weighs too much right now, something on the order of fourteen pounds. Nonetheless, it has some impressive capabilities. Its grenade launcher will reach out at least one thousand meters. It will also have a close range ability for engaging a target behind cover with an air burst round. A laser range finder determines the point at which the munition should detonate. If enemy soldiers are firing from within a building or from a foxhole, our soldiers will be able shoot the air burst munition over or behind the target for a successful engagement. The bursting radius of the round is approximately three meters. We also have a crew-served weapon that's headed in the same direction. The challenge is to keep the cost of rounds at a reasonable level. Currently they are quite expensive. We have a plan in place to cut that cost in half.

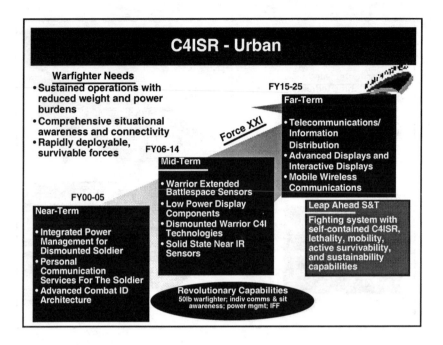

If we recall the challenges confronted by our soldiers as described in *Black Hawk Down*, communications were a fundamental element. The warrior needs to know his location, that of his comrades, and the enemy's positions. The question is, can we borrow from the commercial industry to solve some of these problems? They face some of the same challenges we do, but at the same time our requirements differ somewhat from theirs. We can't take a large communications infrastructure with us, obviously, and we have to be able to function with systems on the move.

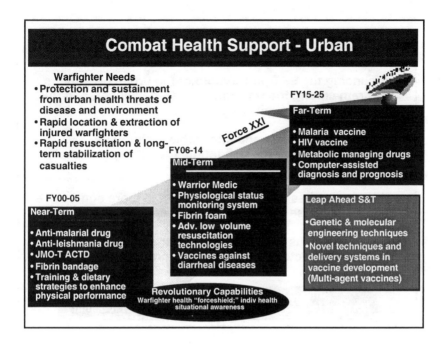

Let's wind up with another piece of the problem: combat health support. We need to be able to monitor the warfighter's physiological status. We have a new ACTD in this area called Joint Medical Operations TeleMedicine. For nonwarfighting scenarios, we want to be able to go in with a minimal footprint and call back to Walter Reed or another facility in order to solve a medical problem. We want to be able to reach back and get as much information as is necessary to save soldier, marine, and noncombatant lives.

Army S&T Bottom Line

- Enhancing the S&T partnership of the urban operations program among the warfighter, user, developer, and the technology community

- Improving the timely and affordable technology transitions leading to warfighting superiority in complex terrain

- Shifting emphasis of S&T investment in more aggressive technology base and more focused on long term

- Increasing focus for leveraging other services, commercial sector, and other government agencies

Striving to Maintain a Balanced & Dynamic Portfolio for Today . . . Tomorrow . . . and the 21st Century

This is our bottom line.

THE ROLE OF AEROSPACE POWER
IN JOINT URBAN OPERATIONS
Maj Gen Norton A. Schwartz, USAF

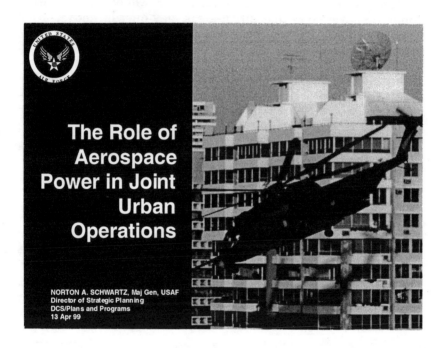

The Role of
Aerospace
Power in Joint
Urban
Operations

NORTON A. SCHWARTZ, Maj Gen, USAF
Director of Strategic Planning
DCS/Plans and Programs
13 Apr 99

Purpose: This briefing highlights the central shaping, controlling, and outcome-determining potential of aerospace power (AP) in joint urban operations. In the context of this briefing, aerospace power includes the following: fixed and rotary-wing systems, land- and sea-based systems, manned and unmanned systems, and space-based systems.

Major theme: The urban environment presents a unique set of challenges for national security decisionmakers and military commanders. Traditional approaches toward military strategy and force employment do not necessarily offer politically and militarily acceptable solutions in the urban arena. Our goal is to examine new concepts for employing joint forces in urban operations that maximize effectiveness and minimize risk.

Main point #1: Successful urban operations will be joint. To overcome the advantages that urban areas may offer our adversaries, CINCs must fully explore the unique capabilities that each service brings to an urban operation.

Main point #2: As a specialized subset of the JFC's overall "tool box," AP provides the JFC with a unique set of options that increases the effectiveness of a full spectrum joint campaign plan while minimizing the risk to forces committed.

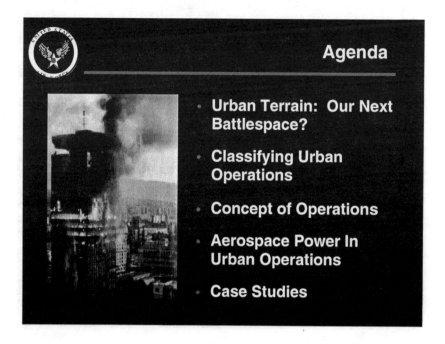

I'll begin with a look at various themes central to understanding urban operations in a strategic context. These include:

- The impetus behind the DoD's heightened interest in urban operations.

- Alternative theoretical approaches to urban military operations.

- The need to define the range of military operations in urban areas.

Next, I will briefly outline an operational concept that might apply in an urban operation. Then I'll show how aerospace power fits into that operational construct. I will conclude with a look at some recent case studies in which AP has played a central role as a "shaper," an "enabler," and an "outcome-determiner" in classic urban operations. Again, our central theme is that effective approaches to urban operations that minimize risk should include aerospace power as an integral component of the joint campaign plan.

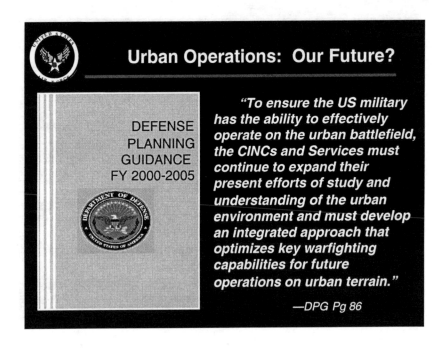

Urban Operations: Our Future?

DEFENSE PLANNING GUIDANCE FY 2000-2005

"To ensure the US military has the ability to effectively operate on the urban battlefield, the CINCs and Services must continue to expand their present efforts of study and understanding of the urban environment and must develop an integrated approach that optimizes key warfighting capabilities for future operations on urban terrain."

—DPG Pg 86

The importance of analyzing, defining, and developing an urban-focused operational capability within our armed forces has received an ever-increasing emphasis over the course of the past decade. Key national security documents, defense panel recommendations, independent analytical studies, joint and unilateral service doctrinal/wargaming initiatives, and practical experiences gained during contingency operations around the globe all indicate that urban operations have held and will continue to hold a central place in U.S. national security strategy development and policy execution.

DPG FY 00-05 states:

> To ensure the U.S. military has the ability to effectively operate on the urban battlefield, the CINCs and services must continue to expand their present efforts of study and understanding of the urban environment and must develop an integrated approach that optimizes key warfighting capabilities for future operations on urban terrain.

National Defense Panel concerns:

- A particularly challenging aspect of the future security environment will be the increasing likelihood of military operations in cities. (NDP Report, p. 15)

- We must also expect to be involved in cities while conducting such contingencies as humanitarian and disaster assistance, peacekeeping, and peace enforcement operations. (NDP Report, p. 16)

- Although we might prefer to avoid urban situations, mission requirements in peace and war may not allow for this preference. (NDP Report, p. 37)

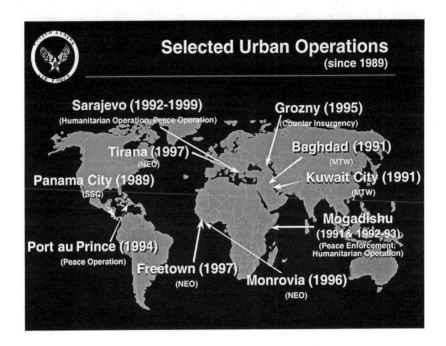

Various futurists indicate that projected trends in Third World urbanization, infrastructure deterioration, ethnic violence, and the like will lead to increased U.S. involvement in urban conflict areas across the globe. Others believe that future adversaries will use urban areas to counter traditional U.S. strengths and lure us onto the urban battlefield.

Whether or not you are a proponent of any or all of these troublesome scenarios, a look back at recent post–Cold War conflicts reveals that military operations in urban environments have been a reality and are here to stay. This graphic depicts a sampling of important conflict zones across the globe where operations in urban environments have held a central role across the continuum of conflict.

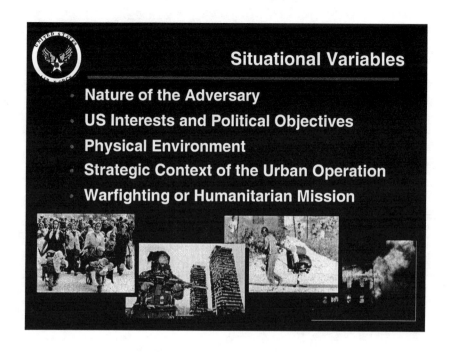

Situational Variables

- Nature of the Adversary
- US Interests and Political Objectives
- Physical Environment
- Strategic Context of the Urban Operation
- Warfighting or Humanitarian Mission

As a conceptual starting point for our discussion of the role of AP in joint urban operations, it is important to acknowledge that urban conflict scenarios are highly diverse in nature. They range in scope and complexity from isolated criminal/terrorist acts committed against U.S. servicemen, through organized urban insurgency bent on national socio-political transformation, to force-on-force conventional conflict at the major theater war level. As such, we cannot afford to focus exclusively on one type or level of urban conflict to the exclusion of others.

When tackling this dilemma, U.S. policymakers and commanders must develop strategies and determine required capabilities and resources based upon in-depth consideration of numerous interrelated factors. These include the nature of the adversarial threat in terms of its size, type, and organization; the geopolitical and physical environments of the scenario at hand; the primacy of U.S. interests and objectives and U.S. domestic support; international/coalition support considerations; and the physical resources that can be mustered at the national/theater level. It is a given that the operational strategy, tactics, and capabilities the United States will bring to bear to win an urban battle during a major theater war will be markedly different from those required to conduct an effective urban counter-guerrilla campaign, an opposed NEO, or peace enforcement operation.

Defining the Urban Threat

	Terrorist/Criminal	Insurgent	Conventional
Organization	Simple; Covert	Variable	Complex; Overt
Political Objective	Variable	Seize Political Control	Control Territory
Military Strategy	Fear	Attack Political Will	Attack Military Force
Resources	Few	Light Arms	Conventional Arms
Popular Support	Minimal	Variable	Variable

To further illustrate this point, let's take a look at one variable in the urban equation: the nature of the threat. While not comprehensive, this slide illustrates the potential complexity of the urban threat across several dimensions. The point here is that the JFC will require a very different set of capabilities and strategies to combat an urban terrorist threat rather than a large-force conventional adversary operating in urban terrain. All too often we fail to make this distinction when talking about urban operations.

Recognizing the varied character of operations that we may conduct in urban areas, the JFC will have to use all the tools available to him, including air, land, and maritime components, to maximize effectiveness and minimize risk.

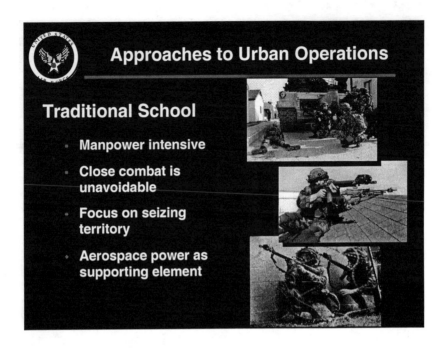

Analysts who study the application of military power in an urban environment agree on one fundamental premise—namely, that military operations in urban environments represent extremely challenging and complex undertakings with overlapping environmental, infrastructure, and population considerations.

From this basic premise, two divergent schools of thought have emerged. The more traditional, land-centric school argues that urban environments offer our adversaries certain distinct asymmetric advantages that, at the same time, largely negate traditional U.S. strengths in the areas of technology and firepower (e.g., AP in all its collective forms to include C4ISR, strike, mobility, etc.). They see urban operations as extremely manpower intensive, as having a focus on seizing/occupying urban terrain via close-quarters infantry-style combat (i.e., house-to-house fighting), and as requiring "low-tech" solutions to battlespace management problems unique to the urban environment.

In this school, service doctrine, academic analyses, and wargaming efforts tend to emphasize the tactical level of war in an MTW context with a specific eye toward small-unit TTP development in the urban environment. The aerospace and maritime aspects of joint military power (including air and maritime superiority) are also largely relegated to a ground support role in this construct.

We acknowledge that there may be some "high-end" scenarios that warrant the political risks and human costs inherent in this approach. However, as most conflicts in urban areas fall short of this mark, this school of thought offers the NCA no realistic, acceptable military options along the remaining conflict spectrum.

Approaches to Urban Operations

Alternative School

- Operational-level approach
- City is a knowable, system of systems
- Nodal analysis identifies key urban components
- Focused on *effects*, not territory

Transportation Networks...

Communications Networks...

High Value Point Targets

A competing school of thought—one that contains much less inherent risk and offers a more universally plausible set of military options—involves understanding the city as a "system of systems," so we can control an adversary without necessarily introducing a large ground combat force. This school proposes that the urban environment comprises key nodes "archived" and susceptible to analysis in considerable detail across three dimensions. By making these key nodes the foci of the joint campaign, we can apply our asymmetrical joint strengths against key adversary centers of gravity without having to close with the enemy in a costly force-on-force confrontation.

This school is more flexible in that it advocates applying a combination of air, ground, and maritime solutions appropriate to the situation at hand and key enemy vulnerabilities. Succinctly speaking, this school targets the shaping and controlling of adversarial behavior through an "operational effects" orientation vice the "territorial" (read "block-to-block") focus inherent in the "ground-centric" approach previously discussed. In the end, I believe that this second school of thought more appropriately lays the groundwork for operational-level successes in the urban environments we must face.

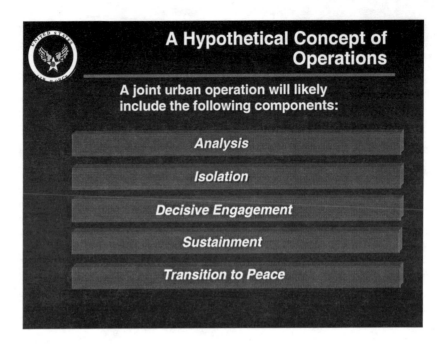

Using the nodal analysis and situational variance concepts advanced previously, let's now discuss a model CONOPS for joint force urban operations. The point is not to argue the merits of this concept over others, but rather to provide a framework to show how AP contributes to the JFC's overall campaign plan.

Most urban operations will include the following components to some degree:

- Analysis
- Isolation
- Decisive engagement
- Sustainment
- Transition to peace

These components should not be viewed as linear or sequential stages in a "by-the-numbers" approach to urban operations. Rather, they should be regarded as interdependent, continuous, and frequently overlapping processes comprising a fluid joint campaign plan for urban operations.

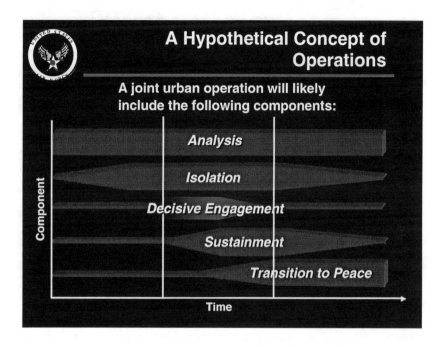

As we discussed earlier, the situational context will, in the end, determine the specific actions, time frames, resource commitments, etc., associated with each component of our CONOPS. Additionally, this slide illustrates that as the urban campaign unfolds, the JFC will most likely shift the weight of effort to the appropriate component to meet the requirements of the situation at hand.

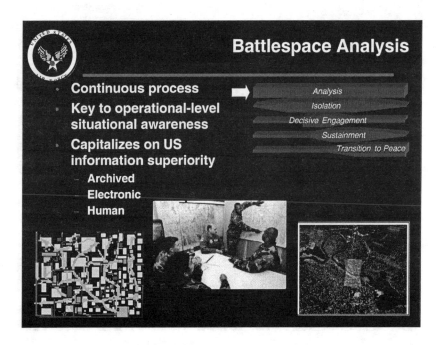

The first, and most important, component of our CONOPS is *bat-tlespace analysis*. This component encompasses intelligence prepa-ration of the battlefield/battlespace, maintenance of operational-level situational awareness in all phases of campaign planning and execution, and it better allows us to predict, compel, and respond to adversary behavior. Proper battlespace analysis allows the JFC to capitalize on information superiority to thoroughly identify/analyze adversary key nodes, systems, dependencies, and choke points cen-tral to his ability to shoot, move, and communicate across three dimensions in the urban environment. Conclusions resulting from our adversary analysis must be based upon a mutually supportive combination of human, electronic, and "archived-derived" data.[1]

---

[1]The term "archived-derived" refers to the nodal analysis premise that cities are made up of interconnected telecommunications, transportation, power systems, etc., that are generally "archived" to some level of detail in various accessible data banks worldwide.

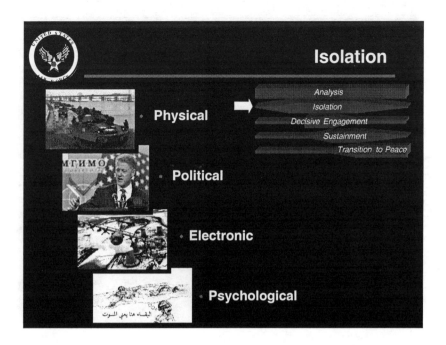

Isolation can take many forms. It implies physical and psychological separation of our adversary from his support base (internal as well as external to the city), denial of his freedom of movement and communication, and negation of his ability to acquire useful intelligence vis-à-vis friendly operations.

For our purposes, isolation also implies physical protection of the urban population from adversary attack and exploitation, as well as the unintentional collateral effects of urban combat. Isolating the enemy while simultaneously affording the appropriate level of protection to the urban population at large requires an active combination of physical, political, electronic, informational, psychological, and civil affairs measures.

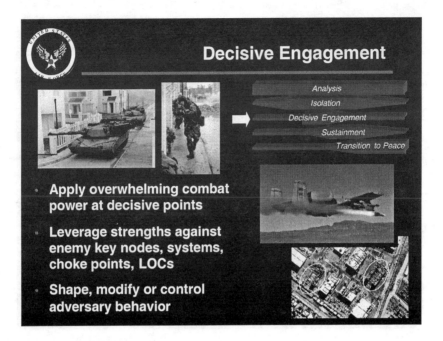

Decisive engagement involves gaining the initiative by applying friendly strengths against the key operational level nodes, systems, choke points, or other targets identified during the JFC's urban battlespace analysis. The ultimate goal here is to weaken or destroy adversary cohesion, organization, command and control, LOCs, and psychological balance so as to ultimately shape, modify, or control his behavior in line with the JFC's campaign plan.

Decisive engagement activities may include precision air strikes against adversary forces/facilities, electronic disruption of adversary communications, insertion of SOF or a conventional ground force to seize a key facility or structure or interdict adversary reinforcements, and others. Continuous reassessment of previous battlespace analysis considerations and maintenance of battlespace situational awareness are paramount to the success of Decisive Engagement operations.

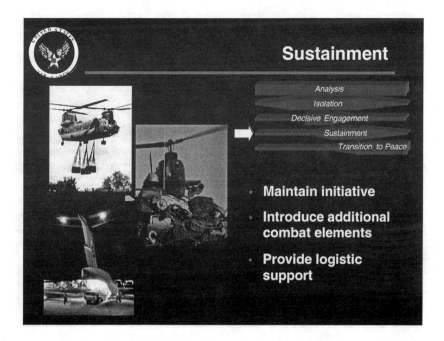

Sustaining the momentum seized in decisive engagement operations is of critical operational-level concern for the JFC. A recurring theme here is the continuous analysis of evolving battlespace awareness, operational effects achieved against key adversary nodes and systems, and adversary responses initiated during decisive engagement operations. Sustainment activities may include any or all of the following:

- Introducing additional elements of power to maintain tempo;

- Providing logistics support to committed forces;

- Achieving cascading effects against remaining key nodes and systems.

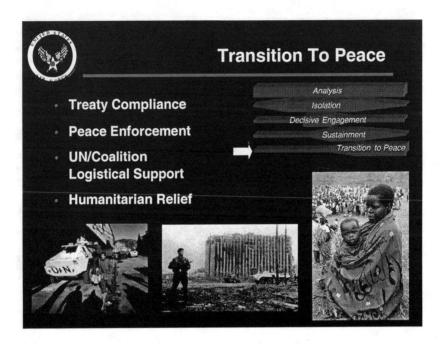

Transition to peace refers to a variety of long-term peace-shaping operations that ultimately mark the termination of urban violence or armed conflict. These may include facilitating the introduction of international peacekeeping or humanitarian relief organizations, setting conditions for a change of regime, destroying conventional or unconventional weapons stockpiles, disarming rival warring factions, and the like. Additional key features of the transition to peace involve a managed transition to a replacement military force or local/international control; the reestablishment of critical urban services; the repair of critical urban infrastructure; and the conduct of follow-on consequence management operations as required.

Now that we've established a hypothetical CONOPS supporting joint force urban operations, let's take a look at the key capabilities that AP offers in support of the JFC's urban campaign plan.

Air superiority: Control of the air is key in that it allows us to maintain the initiative while denying our adversary freedom of action. It also serves as a critical enabler for all other AP operations as well as friendly ground force freedom of maneuver.

Battlespace awareness: Aerospace systems/sensors provide the focused ISR (IMINT, ELINT, SIGINT) and all-source fusion, planning, and decision support paramount to the JFC's continuous three-dimensional urban battlespace analysis process. These systems also provide the information superiority and continuous overwatch and battlespace communications that are the core of effective joint force urban C2/battlespace management.

Precision strike: Precision strikes allow the JFC to conduct multiple, asymmetrically focused attacks against key adversary nodes/systems/LOCs that form the core of his strategy and combat power.

Strike operations may also directly support ground force maneuver and combat operations. Precision strike allows us to achieve a wide range of direct operational-level effects against the adversary while minimizing collateral damage against civilian infrastructure/populace.

Mobility: Aerospace mobility forces enable rapid strategic power projection of U.S. forces worldwide. In the urban setting, tactical lift assets enable high-tempo operations, rapid vertical maneuver, and sustainment of friendly forces.

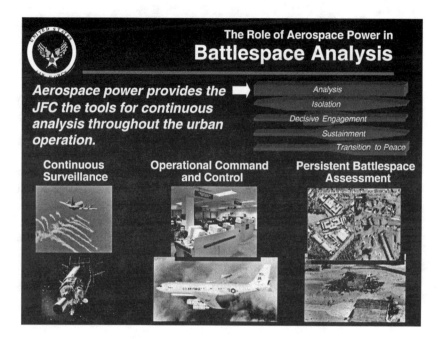

This next series of slides examines the capabilities offered by specific aerospace weapons systems and platforms while describing the roles these systems and platforms play in each of the interdependent processes comprising our urban CONOPS.

AP allows the JFC to better "see" the urban battlespace across three dimensions through a combination of manned and unmanned aerial platforms and space-based systems. These systems seek to provide IMINT, SIGINT, and ELINT and the attendant all-source fusion that, in conjunction with archived data and ground-based HUMINT, gives the JFC a complete picture of the battlespace. These collection systems also contribute significantly to the adversary nodal analysis, campaign planning, decision support, and "sensor-to-shooter" rapid targeting processes. Examples include focused ISR provided by orbital satellites, U-2, E2C Hawkeye, P-3 Orion, OH-58, Global Hawk, Predator, and other systems.

Overhead ISR systems also play a key role in setting the stage to allow the joint team to seize the initiative once hostilities begin. The moving vehicle identification, tracking, and targeting information provided by the wide array of sensors on board overhead platforms such as the JSTARS and AC-130 gunship serve as perfect cases in point.

With advancing technology, airborne ISR systems may also play a key future force protection role. Examples include development of overhead counter-sniper and counter-fire capabilities.

Aerospace systems also provide the dynamic C2 the JFC needs in order to exploit his superior battlespace awareness and conduct proper high-tempo battle management of friendly forces engaged. Robust overhead C2, combined with real-time, three-dimensional intelligence fusion, allows the JFC to effectively operate inside the adversary's decision and action cycle—a key to full spectrum dominance during subsequent combat operations. Examples of aerospace systems providing a wide variety of real-time, multidimensional communications to joint forces include space-based communications satellites, AWACs, P-3 Orion, C2-configured tankers, and lift assets.

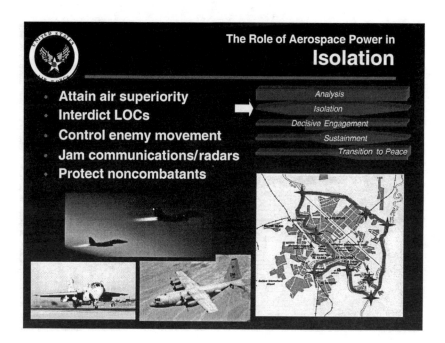

AP can play a critical role in helping establish the appropriate level of information, physical, and psychological isolation over our **adversary**—and, thereby, helping shape his perceptions/behavior and limit his options at all levels before the initiation of hostilities— during an urban campaign.

Synergistic application of selected AP capabilities can also help in the formulation of **population and infrastructure protection** options vital to minimizing noncombatant casualties and physical destruction in an urban setting. Examples include communications/radar jamming, signal overpowering, open frequency transmissions, EC-130 TV broadcasts, leaflet drops, and F-16/F-15/F-14 combat air presence operations.

We have yet to fully explore additional concepts and technologies involving the delivery of nonlethal weapons from the air. This is a logical next step for AP in which we capitalize on gains achieved in weapons accuracy by adding a range of nonlethal effects, many having applications in an urban context (e.g., crowd control).

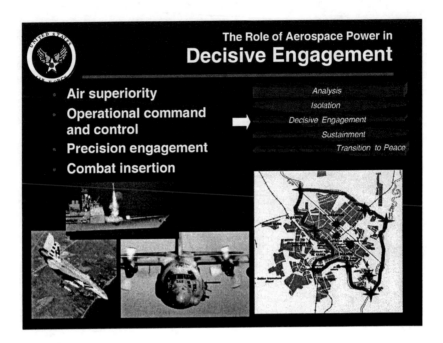

AP plays an absolutely pivotal role in the decisive engagement process. Aerospace forces can bring overwhelming firepower to bear (in the precision strike or interdiction mode) to achieve devastating direct operational-level effects against key adversary nodes identified during the JFC's all-source intelligence fusion process. Examples include B-2, F-117, F-15E, F-16, F/A-18, TLAM, and future spaced-based directed energy weapons strikes.

Future generations of "smart-brilliant" weapons will allow even more precise effects against high-value adversary targets with a low asset-to-target ratio (and, hence, minimize exposure of friendly forces to hostile fire) and minimum destructive effects against noncombatants and civilian infrastructure. Examples include laser-guided PGMs with 4- to 12-pound warheads, MCDWs (F-16 certified), smart fuses to allow kinetic or explosive kills, deep penetration weapons, and next-generation miniature munitions.

Tactical strike assets can serve as key enablers for conventional/SOF ground force operations in the preplanned attack as well as CAS mode. Examples include the AC-130, A-10, AV-8, AH-64 attacks against emergent targets, and adversary reinforcing maneuvers. (NOTE: CITS provides highly accurate GPS-based targeting coordinates from ground FAC to F-16 cockpit computer via data burst, eliminating past reliance on lengthy verbal transmissions.) Tactical lift forces provide the vertical maneuver that allows the pin-point introduction of ground force elements in support of the JFC's overall campaign plan. Some examples are troop insertions via MH-53, CV-22, CH-47, or UH-60.

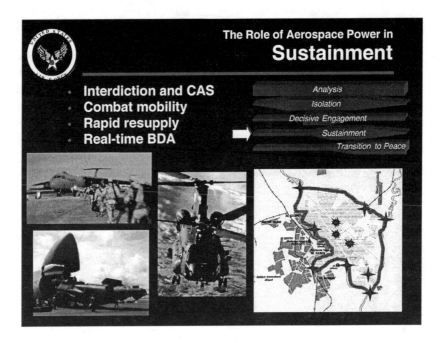

Sustaining the initiative, exploiting effects achieved, maintaining pressure against the adversary, and counteracting adversary moves made during the decisive engagement process are primary prerogatives of the joint aerospace team. Robust overhead C2 and the continuous passage of updated sensor-to-shooter data to the strike assets previously discussed enable the application of cascading second- and third-order operational effects against remaining adversary key nodes. In concert with ground force operations, joint aerospace strike and mobility assets work in tandem to sustain friendly momentum, maneuver and force protection, eliminate pockets of enemy resistance, and resupply forces on the ground while denying the same to our adversary.

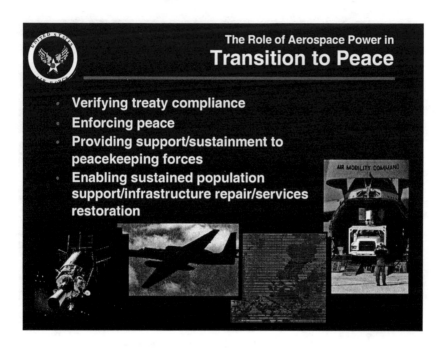

The Role of Aerospace Power in
Transition to Peace

- Verifying treaty compliance
- Enforcing peace
- Providing support/sustainment to peacekeeping forces
- Enabling sustained population support/infrastructure repair/services restoration

Aerospace systems and platforms can serve many critical functions during the transition to peace. Long endurance UAVs, manned ISR platforms, and a combination of continuous multiple coverage satellites can provide sustained "high ground" overwatch of the urban zone as well as provide the sensor-to-shooter data required to support long-term "air constabulary" operations. A wide range of air-to-air and air-to-ground aerospace assets is available to the NCA/JFC to provide the combat punch necessary to enforce adversary peace compliance if required. Enforcement of no-fly zone restrictions and adversary truce line and weapons cantonment violations serve as cases in point.

The strategic and tactical lift capabilities provided by joint aerospace forces are integral to the sustainment of post-conflict international peacekeeping, humanitarian relief, infrastructure repair, services restoration, and consequence management operations. Examples are C-17, C-5, C-130, CH-47, and UH-60 operations supporting food distribution, medical evacuation, international observer movement, and resupply operations.

Having completed our look at how aerospace power contributes to the JFC's urban operational campaign in a generic sense, let's now examine some recent contingency operations that illustrate specific applications of aerospace power in an urban environment. I've chosen the examples of Panama, Somalia, Haiti, and Bosnia due to that fact urban factors were central to the JFC's campaign plan and conflict resolution in each case.

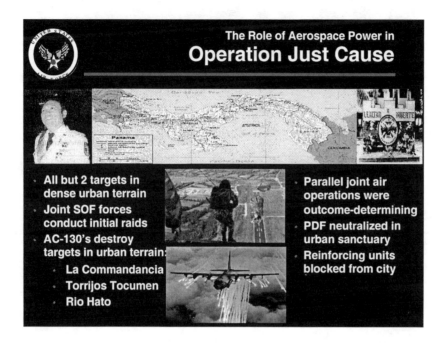

JUST CAUSE was, at the time, the largest and most complex air operation conducted by joint U.S. aerospace forces since the Vietnam era.
The initial joint forced entry operation and follow-on search for General Noriega and his cronies involved extensive urban combat and
civil pacification operations. Aerospace forces played a critical role
in protecting U.S. citizens and defeating PDF elements seeking
refuge in urban terrain. The following parallel joint aerospace
operations were key to outcome determination across the JFC's
campaign plan.

- H-hour assault plan included 140 aircraft (representing 16 different types of fixed/rotary assets) participating in more than 250
  sorties in built-up areas.

- AC-130s conducted precision strike operations and supported
  SOF urban operations throughout the country. AC-130s also de-
  stroyed the 15-building Commandancia complex in downtown
  Panama City, crippling the single most important node in the
  PDF national defense/C2 systems.

- MC-130s, C-130s, C-141s, and C-5s conducted strategic airdrop/airland operations direct from CONUS onto built-up area Torrijos-Tocumen/Rio Hato airfields.

- EC-130s jammed commercial radio/TV stations and PDF radio C2 nets.

- Army MH-47/MH-60s provided critical vertical maneuver for ground forces.

- Aerospace power neutralized PDF units in their urban sanctuary and interdicted key reinforcing units at numerous chokepoints throughout the city.

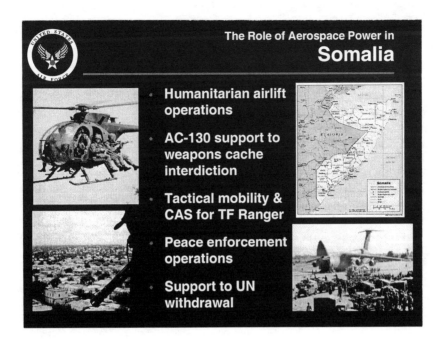

From initial entry operations supporting RESTORE HOPE in 1992, through UNOSOM II in 1993, to the eventual withdrawal of UN forces during UNITED SHIELD in 1995, Somalia represented one of the most airpower-intensive operations in the post–Cold War era.

U.S. strategic/tactical lift assets accomplished the lion's share of the initial humanitarian relief operations, with the first U.S. strategic airlifter touching down four hours after the Marines secured Mogadishu International Airport. (Over 6,000 airlift sorties were eventually flown in support of the three-year relief/peacekeeping effort in Somalia.)

Army lift assets provided critical vertical mobility to UN urban quick-reaction forces and established an effective "air presence" over Mogadishu to compel desirable warlord and militia behavior at various key times. Army helicopter gunships provided continuous, critical precision fire support (to within 15 meters danger close distance of friendlies) and overhead reconnaissance and armored relief convoy escort, saving the day for a beleaguered Task Force Ranger in the longest sustained firefight involving U.S. ground forces since the Vietnam War (eighteen hours). In fact, had AC-130 gunship support been available in-theater for the final TF Ranger operation against fugitive warlord Aideed, events of that fateful day would most likely have taken a completely different direction, and with minimum U.S. losses.

Following the 3 October 1993 TF Ranger operation, U.S. strategic lift brought significant reinforcements (1,791 troops, 3,721 short tons of critical cargo, 9 M-1 tanks, and 14 Bradley FVs) to Mogadishu in less than 72 hours. AC-130 gunships successfully conducted "air presence" deterrent operations, targeted illegal militia weapons caches, and provided deadly accurate close air support to UN ground forces during the final days of the UN withdrawal in a steadily deteriorating urban security environment.

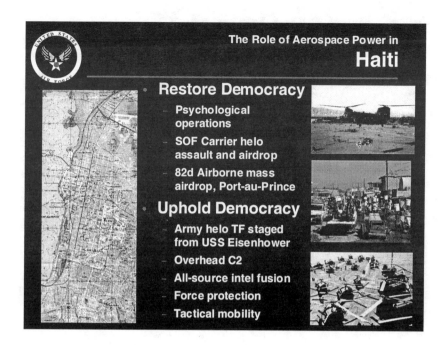

Operation RESTORE DEMOCRACY illustrates aerospace power's ability to help "isolate" an adversary in urban terrain. In preparation for the introduction of U.S. forces into Haiti, EC-130 aircraft flew preintervention psychological operations (PSYOP) broadcast sorties and were prepared to disrupt command and control. In addition, MC-130s conducted leaflet drops over Port-au-Prince to educate the civilian populace on U.S. intentions and break the fighting spirit of government security forces.

In terms of "decisive engagement," RESTORE DEMOCRACY originally envisioned the near-simultaneous forcible entry of two independent task forces. Had combat operations been necessary, one such force would have executed a helicopter assault from an offshore carrier against a half-dozen key security nodes in the capital area. These men would have been supported by an armada of fixed and rotary-wing close air support assets.

In recognition of their precision capability, AC-130s received approval for preplanned strikes against heavy weapons storage facilities. Moments after this operation was under way, a conventional joint task force was to seize the airport and areas north of the city via airdrop, air landings, and long-range rotary wing insertions direct from the United States.

The operation from the USS *America,* involving almost 100 air assets of all types, was a first in terms of joint cooperation.

The air plan supporting the conventional joint task force, with its requirement to rapidly deliver three brigades of the 82nd Airborne Division, would have been the largest in scope and complexity since MARKET GARDEN in 1944.

The scope of this air armada may well have prompted the Haitian strongman Raoul Cedras to accept a peaceful transfer of power.

After Cedras' departure, aerospace forces provided robust overhead command and control, tactical lift, theater search and rescue, and protective overwatch for the international force, all key aspects of "sustainment" and "transition to peace" missions.

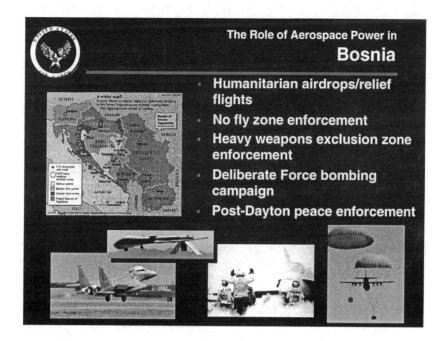

The Bosnia case study represents the "Crown Jewel" example for highlighting the outcome-determining potential of AP in an urban environment.

U.S. joint urban aerospace operations supporting UN/NATO efforts in Bosnia can be divided into four broad categories: humanitarian airdrops, relief flights, and evacuations; enforcement of the UN-mandated no-fly zone; enforcement of the heavy weapons exclusion zone around Sarajevo; and provision of critical overhead ISR of Sarajevo and the surrounding region.

- 160,000 Mtons of humanitarian assistance were airdropped/airlifted to isolated urban areas; this accounted for 85 percent of the international relief brought into Sarajevo.

- The credible threat of NATO air strikes compelled Serb forces to withdraw 300 heavy weapons from Sarajevo in February 1994, with another 290 transferred to UN control.

- F-16s shot down four Serb Galeb fighters while enforcing the UN no-fly zone in February 1994; this incident "shaped" Serb behavior/fixed allied air supremacy.

- DELIBERATE FORCE, conducted in the summer of 1995, illustrated the successful combination of integrated C4ISR, rapid information processing/sharing, and long-range precision strike against a target set in which 75 percent of the targets struck were in urban terrain. Nearly 99.5 percent of aerial munitions employed were PGMs, producing almost zero collateral damage. Operational effects achieved seriously altered Bosnian Serb military superiority vis-à-vis Muslim opponents and weakened the Serb leadership's willingness to continue the fight, ultimately delivering the Serbs to the Dayton Peace Process.

- Post-Dayton: U-2s and space-based surveillance assets monitored treaty compliance, helped locate mass burial sites, and delineated the permanent ZOS line between former warring factions. JSTARS provided overwatch for NATO ground operations, as well as tracking of Serbian maneuver forces and monitoring of numerous ZOS activities. NATO fighters/attack helos provided continuous air presence over disputed areas, helped quell spontaneous civil disturbances, and enforced tactical-level treaty compliance and integrity of the inter-entity boundary line.

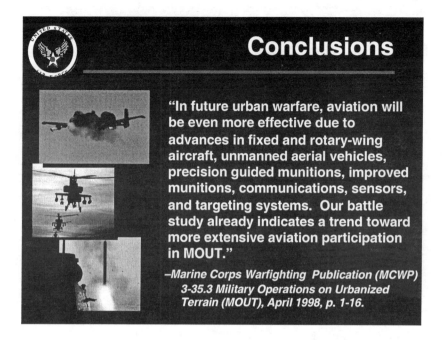

Conclusions

"In future urban warfare, aviation will be even more effective due to advances in fixed and rotary-wing aircraft, unmanned aerial vehicles, precision guided munitions, improved munitions, communications, sensors, and targeting systems. Our battle study already indicates a trend toward more extensive aviation participation in MOUT."

–*Marine Corps Warfighting Publication (MCWP) 3-35.3 Military Operations on Urbanized Terrain (MOUT), April 1998, p. 1-16.*

The case studies I've just reviewed bring up numerous key points that I need to highlight once again before concluding this briefing. First and foremost among these is the fact that AP has historically played a decisive role in joint urban operations at the strategic, operational, and tactical levels. With the emergence of next-generation air- and space-based "eyes and ears," enhanced overhead C2 systems, robust strategic/tactical mobility, and precision strike MCDWs, AP will continue to offer the NCA/JFC a greatly expanded range of cross-spectrum options to shape, control, or decisively defeat an urban adversary well into the future.

AP is integral to any joint concept that will offer the JFC an effective military strategy while minimizing risk. Whether in the context of achieving direct, precision operational effects against key adversary nodes; decisively enabling high-tempo parallel joint force operations; or providing real-time, three-dimensional, actionable ISR to the JFC; AP commands an absolutely pivotal position in the joint force urban campaign plan. Failure to bring the advantages inherent in joint aerospace power to bear against our adversaries in the urban environment puts operational success seriously at risk. A full complement of cross-spectrum joint power is the key to achieving our national objectives in this most challenging of all operational environments.

MOUT: A DOMESTIC CASE STUDY—
THE 1992 LOS ANGELES RIOTS
MG James Delk, CAARNG (Ret.)

Editor's note: General Delk's presentation began with a videotape clip of commercial television news segments, announcing the results of the trial in Simi Valley, California, in which Los Angeles policemen were found not guilty of assaulting Rodney King. This was followed by amateur video camera footage taken by a security guard in Los Angeles, who was also a military policeman in the California Army National Guard. That footage included twenty-seven gunshots fired by those *participating in the riots. Some of those shots were aimed at fire engines and ambulances. The final clip showed looting the next morning.*

General Delk: This photograph was taken about 11:00 in the morn-ing, and as you can see, the smoke over Los Angeles almost turned day into night. That highlights one of the many differences we found regarding this riot as compared to those of the 1960s. Do you remember the riots affiliated with the Vietnam War protests? The riots in Detroit, Watts, and Newark were very clearly nighttime phe-nomena. The cops, soldiers, looters, and rioters rested during the daytime.

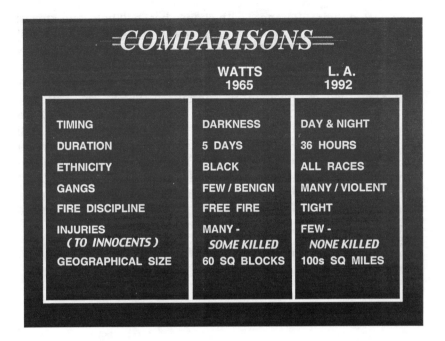

This slide lists some of the differences we found between the 1992 Los Angeles riots and those of earlier years. There are many, as you can see. The first one I have already mentioned, the difference between the times the riots occurred. That had a big impact. For instance, the police knew when the trial would be over. The LAPD's SWAT team was told come to work at 6:00 P.M. because previous history had demonstrated that people don't riot during the daytime. The police therefore had to be ready for deployment in the evening so as to be ready for action when it got dark. Well, by the time they reported for work, the riots were already under way.

Duration was a second difference. The 1965 Watts riots were five days long, while for all practical intents and purposes, the 1992 riots were over in 36 hours. Another big difference was the ethnicity. You get the impression from the early film clips that it was a black/white riot, which was very typical of those in the 1960s. Not so in 1992. As a matter of fact, most of the arrests were Hispanics, many of whom were illegal immigrants. Further, there were not nearly as many gangs in the 1965 riots. Gangs in earlier years were rather benign.

They settled their differences with chains, baseball bats, and knives; guns were comparatively rare. In 1992 they had literally thousands of guns, many of them better than ours. In addition, they very quickly looted gun warehouses and ended up with many more weapons. Another difference related to the fire discipline demonstrated by military and law enforcement personnel. In 1965 it was "Katie bar the door." We had .50 caliber machine guns and we used them. If a sniper shot from behind a concrete wall, we just opened up with a .50 and cut the wall down. If he sniped at us from a rooftop, we cut off the top of the roof. There were problems with that of course; we suspect we killed at least five innocent civilians. We were very aware of that in 1992. Sergeants major and senior officers remembered that we had killed innocent people in 1965. It had a big impact on the rules of engagement and how we handled ourselves.

Geographical size—you can see the Watts riots covered 60 square blocks. In 1992 it was literally hundreds of square miles.

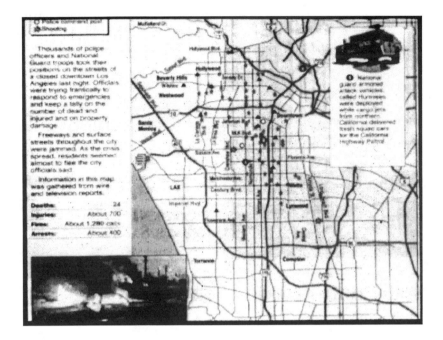

There's a map. From the top to the bottom of this map is 32 miles, and this reflects only the area covered by actions on the first night of the 1992 riots. And then it goes way off the map to Pomona, covering literally hundreds of square miles, and that's just LA. You may remember that these riots affected several other cities in the United States and Canada. There are a couple of notable points to be emphasized. The first deals with command and control measures. Operational types, the minute they get on the scene, look at easily identifiable terrain features and allocate areas of responsibility. For example, unit boundaries could be drawn along the big freeways because they're so recognizable. That's the automatic reaction of any good operator. Wrong! Los Angeles police commands are not divided up by freeways. Points such as these are very important to understand because the military providing support in Los Angeles had to break down and organize to work with law enforcement. Up here in the corner, you see a picture of a Humvee. The people who labeled this picture made a mistake which worked hugely to our advantage. The caption stated that Humvees were armored vehicles.

They're no more armored than a good Buick in most cases. Fortunately, the rioters didn't know that. We never did have a bullet fired at a Humvee.

I want to also dispel a myth. There are a lot of black soldiers in the 40th Division. One book written by a FBI agent talked about our having to identify gang members who wore America's uniform so as to send them home. If you'll pardon the expression, that's B.S. We know many of our soldiers are members of gangs. We had some problems before the riots with some individuals wearing their gang colors, blue for the Crips and red for the Bloods. That was quickly taken care of by our sergeants major. When those men subsequently put on America's uniform, they were good soldiers. They helped us to understand gang behavior. They were soldiers first and gang members second.

=CHRONOLOGY=

**WEDNESDAY,
29 APRIL 1992**

1245	NOTICE GIVEN - VERDICT IN 2 HOURS
1510	VERDICTS ANNOUNCED
1520	CROWDS GATHER AT NORMANDIE AVE.
1600	MAYOR EXPRESSES SHOCK
1615	LOOTING OCCURS
1745	LT. MOULIN DIRECTS NO FURTHER DISPATCHES TO NORMANDIE
1825	CHIEF GATES GOES TO FUND RAISER

Let me take you through the chronology. At 12:45 the judge told the police that the verdict would be delivered in two hours. They actually had more time than that. Nevertheless, within 20 minutes of the verdict, a crowd gathered at Normandie, the point at which it initially blew up. You heard the mayor's comments about "renegade cops." I don't think it's coincidental that 15 minutes later, looting occurred. You remember that there were some problems with the police department, they'd be the first to admit the terrible mistakes that they made. They didn't realize how serious this initial event was. They described the evening's events as a "Mardi Gras" atmosphere, and in places it was. However, there was a lot of killing going on also. In fact, Chief Daryl Gates went to a fund-raiser.

CHRONOLOGY

WEDNESDAY, 29 APRIL 1992

1845	REGINALD DENNY BEATEN
	CITY - WIDE TACTICAL ALERT CALLED
1900	PARKER CENTER STORMED (DISPERSED ONE HOUR LATER)
2015	FIRST FATALITY: YOUTH SHOT AT VERNON & VERMONT
2045	FIRST REPORT OF ARSON
2100	GOVERNOR DIRECTS MOBILIZATION OF 2,000 GUARD MEMBERS

It wasn't until 6:45 P.M. that a citywide tactical alert was called. The first fatality occurred at 8:15 P.M. during a drug bust that went bad. And at 9:00 P.M. the governor asked for the Guard.

CHRONOLOGY

THURSDAY, 30 APRIL 1992

0042	LOCAL AUTHORITIES REPORT NO NEED FOR NATIONAL GUARD
0300	2,000 GUARDSMEN IN ARMORIES
1007	OES : DON'T EXPECT MISSIONS PRIOR TO 1600
1100	MFC & CHP COMM. / CHP ARRIVE LOS ALAMITOS
1200	FIRST INFORMAL REQUEST FOR TROOPS
1331	FIRST TROOPS READY *(MISSION CHANGES 4X)*

A little after midnight they said they didn't need us even though we were already rolling troops. The reason they thought they didn't need us was because law enforcement intelligence was lousy; they believed it truly was a Mardi Gras atmosphere. There were already people being killed, but they thought they could handle it. We got a phone call from the Office of Emergency Services in California at 10:07 the following morning telling us we weren't needed until that evening. Why? Because nobody ever riots during the daytime! We were all preparing for the "last" war like we always do. That's why I keep pushing the significance of the differences between the 1992 and earlier riots. As he was calling us saying we're not going to need you until tonight, his city was burning down around him. Looting was going on and people were being killed. At 11:00 A.M. I was flown in as what we call the Military Field Commander. Maury Hannigan, you may remember him from the old CHIPs program on television, flew down with me. He was Commissioner of the Highway Patrol. Well, we stepped off the airplane and the first thing I asked was what any one of you would ask: "Who's in charge?" And they didn't know!

This is 11:00 the next morning and the riots had started during the afternoon of the previous day! Nobody knew who was in charge. Now that's cause for any good soldier to have concerns. I did. Maury sensed that; he said "Jim, I'm going to run off and try to get that put together, and I'll get back to you."

So I went off to the division to get briefed on what was going on. By noon I was getting panicky calls from the Sheriff's Department. They wanted to get some soldiers on the street. They knew they were going to have media helicopters flying overhead and this was an opportunity to let the people know that the Guard was on the ground. Well, I had some problems with that; we hadn't had enough time to do all the things we needed to do. Nobody should ever commit to providing soldiers on the street in less than 24 hours if there's going to be shooting. You have to have rules of engagement and arming orders; you have to coordinate unit orders and understand where you go to get your support. All are very important things that keep your soldiers from being killed and are essential to keep from killing innocent people.

We had our first troops ready nonetheless. I emphasize we still didn't know who was in charge. But we were old friends talking to each other, in this case, Undersheriff Bob Edmonds and I. He requested some soldiers so as to be able to at least get the word out to the press. Well, by 1:30 P.M. we had our MPs ready. They're trained for this. Not only that, but I'd say a good third of our military police in the Guard are cops anyhow, so we were very comfortable turning them loose even though we didn't know who was in charge. The mission changed four times, a good example of the so-called "fog of war." You get used to this kind of thing. Every soldier, marine, sailor, and airman gets used to this. Things were happening so fast that leaders didn't know where it was most important to send troops to keep people from getting killed.

CHRONOLOGY

THURSDAY, 30 APRIL 1992

0042	LOCAL AUTHORITIES REPORT NO NEED FOR NATIONAL GUARD
0300	2,000 GUARDSMEN IN ARMORIES
1007	OES : DON'T EXPECT MISSIONS PRIOR TO 1600
1100	MFC & CHP COMM. / CHP ARRIVE LOS ALAMITOS
1200	FIRST INFORMAL REQUEST FOR TROOPS
1331	FIRST TROOPS READY *(MISSION CHANGES 4X)*

Ammunition was arriving in Los Alamitos for our troops. We had given the MPs our drug enforcement ammo, which we keep stocked for that mission. In the meantime, Maury says we're going to get together and agree on who's running this show, and what the flow of command is going to be. So we agreed that we would get together at 2:00 P.M. in the Sheriff's office. We flew over there in a helicopter. Everything in Los Angeles is done by helicopters landing on top of buildings, it's just a fact of life. When we got together there was the Highway Patrol, the Police Department, the Sheriff's Department, and us. We no sooner had sat down and started getting organized than we got a conference call from the governor asking "Why weren't the troops on the street?" There were some interesting discussions there. In any event, in the meantime, the Sheriff's Department took the first troops while we're still meeting to decide how we're going to run all this. Sheriff Sherman Block [head of the Los Angeles County Sheriff's Department] was sitting at the head of the table. He is the number one law enforcement officer in his county. Sitting to his right was his undersheriff, Bob Edmonds. Next to him was Chief

Daryl Gates of the Los Angeles Police Department. I was on the other side of Sheriff Block, and then the Commissioner of the Highway Patrol and a couple of his assistants were next to him. There was a lot of tension in the room. The mayor had not been talking to his own police chief for thirteen months. There was a lot of competition between the Sheriff's Department (county) and the Police Department (City of Los Angeles). They also hadn't been talking to each other for months. In any event, it finally did get going. The first issue decided was raised by Chief Gates. The Police Department had agreed to protect the Fire Department, but so far had been unable to do so. There had been some serious injuries to firemen. Gates looked across at Maury Hannigan and asked the Highway Patrol to take that mission. Maury said "Sure." Then the undersheriff looked at me and asked whether the Guard could take the rest of the missions. And I said "Sure." Then suddenly it slowed down when it was time to agree who was going to run the show. Finally, Edmonds looked at both the Chief of Police and the Sheriff and said it should be run out of the Sheriff's emergency operations center. The Office of Emergency Services, the military, and the various law enforcement agencies were all to have representatives at that facility and all operations would be run from there. We agreed and all went running out. That took us to 3:00 P.M., almost precisely 24 hours after the riots had started.

This is Biscaluz Center, the emergency operations center. The EOC is under ground. You can see all the antennas and what all above it.

These were the first soldiers on the scene. They are military police-man. That's a Winchell's Donut House burning up.

I'm going to try and give you a sense of what was going on in the streets. This used to be a market. They burned out all their markets. Only problem with that was that there was almost no place to buy food in all of south central Los Angeles after the riots.

This is an abandoned parking lot with looted, abandoned cars. You can see that people just jumped out of their cars and started running.

That's a furniture store.

You can tell that this is south central LA. There's grating over the windows. You can see the gang graffiti. By the way, stay on the freeways in Los Angeles. I was born here. I know the town. If you get off the freeway for any reason and you see a lot of graffiti, get back on the freeway very quickly.

These are all little shops with an apartment building in the background.

This is my favorite photo. Rental companies did a brisk business in trucks and trailers when the looting started. That's a rental truck whose driver tried to drive in to steal from that furniture store. The truck is stuck there for good, and was damaged when the building burned down.

You heard that the agreement was that the Highway Patrol would protect the Fire Department, but very quickly it became obvious that firemen wanted rifles around because of the sniping. There's only so much you can do with pistols and shotguns. And so they said they'd appreciate it if we sent along a few soldiers with what they called "long sticks," and so we did. That stopped the sniping. I would point out that I know these guys never fired a round. But just the fact they had rifles and the bad guys knew it made all the difference in the world.

This slide lists the arrests by crime category by day. This is from the Webster Report. William H. Webster was formerly head of the FBI and a federal judge. His team researched and published the definitive report afterwards. You can see that property crimes nose-dived while the curfew crimes went up. You can also see that the riots ended very quickly.

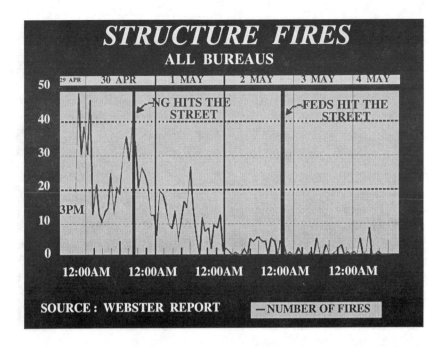

Let's take a look at the fires. And I'm going to make a point here. The National Guard hit the street; as you can see, the fires and crimes quickly ended. This was before the federal troops arrived. Now when I say the fires and the crimes ended, I mean they dropped to about 40 percent of normal. There is normally a lot of gunfire on the streets of Los Angeles. I guarantee that if you went to Los Angeles tonight, you would hear gunfire. There are a lot of bullets flying around. That's just a fact of life. Los Angeles has three times the national average in killings by gunfire, and there are several every night.

CHRONOLOGY

FRIDAY, 1 MAY 1992

0415 ARMY'S 7th DIVISION CALLED TO RIOTS

0600 USMC CALLED TO RIOTS

0900 2,000 GUARD ON THE STREET
      - 2,500 IN RESERVE

1630 JTF - LA COMMANDER LANDS IN LOS ALAMITOS

      1st CHALK - 7th DIVISION HAS WHEELS UP

1900 GOVERNOR REQUESTS FEDERAL TROOPS
      (REQUEST DENIED)

2000 CNG HAS 6,000 ON THE STREET OR STAGED

They called in the Army's 7th Infantry Division. We are fortunate to have Brigadier General Ed Buckley with us. He was then Colonel Buckley, commanding the Second Brigade of the 7th Infantry Division (Light) out of Fort Ord. The marines were also called at 6:00 A.M. In the meantime, we still had a lot of people in reserve. This was frustrating for everybody. That afternoon, the Joint Task Force—Los Angeles commander, Major General Marvin L. Covault, landed. He was a War College classmate of mine, so that helped things. About the same time, the first chalk of the 7th Infantry Division was wheels up. At 7:00 P.M. the governor requested federal troops. We couldn't get them to him because they weren't there yet. It takes time for things to happen. By 8:00 P.M. that night we had 6,000 National Guard personnel on the streets or staging.

CHRONOLOGY

SATURDAY, 2 MAY 1992

0325	JTF AMMO ORGANIZED
0500	LAST OF 7th DIVISION ARRIVES
1040	EL TORO ISSUES RIOT GEAR
1700	USMC ON THE STREET (A + 36:45)
2335	FIRST 7th ID TF CROSSES SP

Early the next morning, ammunition was organized and the last of the 7th arrived and was issued riot gear. The marines were deployed at 5:00 P.M. after being issued riot gear and receiving refresher training. The 7th Infantry Division was initially being held in reserve because we were having trouble finding missions for people. We finally agreed to pull Guardsmen off the street and put some federal troops in their place.

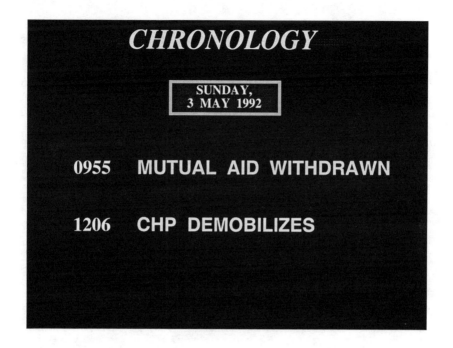

CHRONOLOGY

**SUNDAY,
3 MAY 1992**

0955   MUTUAL AID WITHDRAWN

1206   CHP DEMOBILIZES

Now, interestingly, on Sunday they withdrew the mutual aid. Mutual aid is a civilian system for getting either fire department help or law enforcement help. It's used all over California. I can't tell you about the rest of the nation. But the first thing that's supposed to happen is, you call your fellow cops and they come to help, with the military being the last in and the first out. In this riot, it was reversed: the military was the first in and the last out. They didn't call mutual aid until after we were there. Then they let law enforcement and firemen from other areas go home while our troops stayed on the streets for three more weeks.

SPEED OF RESPONSE

CAL NG	18 HRS	COMMITTED
82d ABN DIV	18 HRS	1st ACFT "WHEELS UP"
GARDEN PLOT	24 HRS	TROOPS STAGED
JTF - LA	36 HRS	COMMITTED

Let's talk about speed of response. We were committed at 1:31 P.M. That's fudging a little bit, as that's just the one MP company I told you about. The rest of the troops wouldn't get there until about 24 hours had elapsed. Just to give you a sense for what the standards are, the 82nd Airborne Division's Readiness Standing Operating Procedure (RSOP) says that they will have their ready brigade's first aircraft wheels up in 18 hours. Then there's page after page after page of what our most ready division has to be doing to get ready to go. You can't believe what goes on in the supply room alone. The federal standard in GARDEN PLOT (the federal response plan) says give the military 24 hours to be staged and 36 hours to be on the street. The federal troops met that standard in Los Angeles.

The National Guard had a secret advantage: We had our own air force (the California Air National Guard) that flew 49 C-130 sorties on the first day. As I pointed out, the 7th didn't get the C-141's they immediately needed to fly them south. They're not like the 82nd Airborne with an Air Force Base right next door. So it wasn't until the afternoon of the day they were called that they even saw their first C-141.

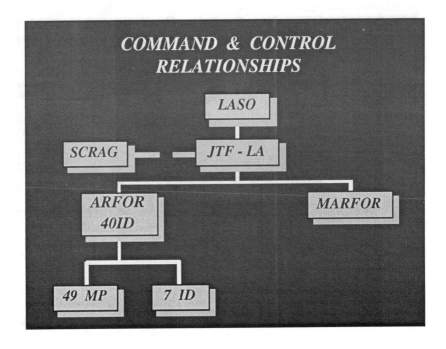

These are the command and control relationships as they existed after we were federalized. This is the standard, out of the manual. It goes from whatever the civilian agency is, in this case the Los Angeles Sheriff's Office, through the military, which was Joint Task Force—Los Angeles. The SCRAG is the Senior Civilian Representative of the Attorney General. He represents the President. You notice most of the lines are solid. However, the line from JTF-LA to the SCRAG is broken; that's filled with lawyers. Suddenly the system was constipated. Everything was going great right up until then. You need to understand what happened after federalization, when lawyers had to carefully review all of the mission requests. This is so important to understanding what happened then, and what, I guarantee, will happen in the future. Missions started taking at least 8 hours to be approved.

The ARFOR (Army Forces) commander was the commander of the 40th Division, Major General Dan Hernandez. The 7th Infantry Division's brigade was commanded by General Buckley, a colonel at that time. And there was a MP brigade. We probably had 13,000 or so Army troops on the scene. The MARFOR (Marine Forces) out of Pendleton represented slightly more than 1,500 additional troops commanded by Brigadier General Ted Hopgood.

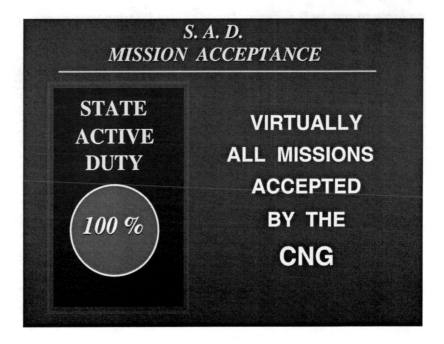

Until federalization, virtually 100 percent of requested missions had been accepted.

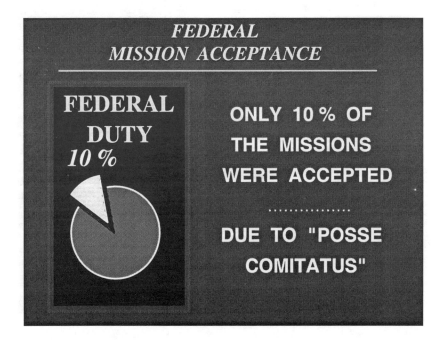

Suddenly it was down to 10 percent. I put *posse comitatus* in quotation marks, because people were saying it was because of *posse comitatus*. Not so. *Posse comitatus,* which has its origins in a federal law passed after the Civil War, says in effect that federal soldiers are not to be put on the streets enforcing laws against civilians. The law was passed because of some excesses following the Civil War. That wasn't the issue in Los Angeles. What had happened in 1992 is that *posse comitatus* had been waived. General Covault knew full well, as did many of his soldiers, that it didn't apply. I'll show you in a minute what did apply.

Soldiers were having trouble with it; law enforcement was having trouble with it. We sat some people down and asked them to put together a piece of paper that everybody could understand. You see a G3 plans officer here, a major. He represented the division. The colonel with the glasses is also a high-level detective in the LAPD; he's both a soldier and a cop. This big guy, Dick Odenthal, is now a captain in the Sheriff's Department. He also was one of the finest intelligence officers in the National Guard that I ever saw, a very, very bright guy. A great soldier, a great cop. Obviously he had the confidence of both sides. He kind of ran the team. This other fellow is a lawyer, but you might be able to see that he is wearing Armor brass. We wanted someone who knew how to say yes, not no. These men sat down and worked the issues out. The product they put out was one of the most important pieces of paper to come out of the LA riots. We handed it to everybody and they suddenly said "Bingo!" The lights went on. The product outlined what were appropriate missions and inappropriate missions depending on the status of the military forces supporting law enforcement. (The Mission Tasking

Guidelines and Rules of Engagement are included as annexes to this appendix.) We had to adjust missions on occasion . The space used to say "Haul Prisoners." The National Guard did that just once. We had a convoy hauling prisoners from one of the housing projects, and all the mothers were standing on the side of the street crying because their "babies" were being carried off to jail by those nasty soldiers. We said let the cops do their own dirty work; that's one we're not going to do for them in the future. We did it once and the California Guard will never do that again. You look at these missions and you see what's appropriate and what's inappropriate. You can see that when you're in state status, you can do almost anything that you both agree is a good mission. You also see, however, on the federal side, that when you're restoring law and order, which was the mission given to the 7th ID and the Marines, that they could also do all the same things. What was the problem? The problem was when they got there, law and order had *already* been restored! So they were stuck over here in the so-called "preserve law and order" status. And as you can see, there's not much they really could do. And that's why these lawyers were denying most of the mission requests. Well the people of LA loved having us there. As I said, the crime rate dropped to 40 percent of normal. We had men come up to our soldiers and tell them how much they appreciated their being on the street. For the first time in years their wives could safely walk to the grocery store. That's also why it took weeks to finally get our soldiers out of there. They simply didn't want to let us go.

FEDERALIZATION DIFFERENCES

- PAY
- CONTRACTING
- LOGISTICS
- ROE / AO
- TASKINGS (POSSE COMITATUS)
- RESTRAINT
- RISK TAKING

INTANGIBLES

There are important differences between state and federal status. I just want to touch on a couple of them. One is pay. In California, for instance, our Guardsman are never paid less than Sergeant E-5 when we call them up for state active duty. Why? Unlike federal troops, their wives can't go off to the military hospital; their wives can't use the PXs or BXs and all those kinds of things that are standard benefits for the federal soldier. And because of the disruption to their lives, we just don't pay them any less than an E-5. What happened? The minute they are federalized, PFCs get paid as PFCs, and not only that, but the changing of the pay systems back again after several days was very disruptive. We had a lot of guys get into real financial difficulties. So that's one of the key significant differences.

Restraint was another area of difference. I used to talk about soldiers and marines being trained to kill, kill, kill. And I can remember when I took basic and went through the bayonet course, you were told to holler "Kill, kill, kill." That's how we were trained. We expect it of our soldiers and our marines. Thank God they're ready if they're called. At the time of the 1992 riots, we weren't pulling all of the peacekeeping missions we're getting nowadays. Now our soldiers and marines are better trained to handle missions calling for restraint, so this difference is not nearly as important as it was at the time.

It's a dangerous environment. I've already talked to you about how dangerous it was in Los Angeles. There were 771 gang-related deaths in 1991. Many more people lost their lives that year in non-gang killings in a city with three times the national average of shooting deaths. There were 102,000 gang members at the time of 1992 riots, and the last time I asked that number had grown to 150,000. And many of them are very, very well armed. We were very conscious of that when we were going in.

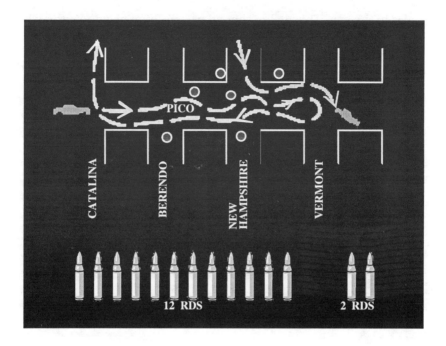

Editor's Note: General Delk ran a video clip from a Los Angeles news account describing an attempt by a civilian to run over several California National Guardsmen. The general identified several errors in the news report.

At this point, this is what happened at Pico Blvd. These personnel were from the 40th Support Company, male and female, trying to enforce curfew. A fellow by the name of Victor Rivas, driving a 240, not a 280ZX Datsun, came down Pico Boulevard going back and forth trying to run over our soldiers. He did a wheelie and headed back trying to run over soldiers as they all hollered "Halt, halt, halt." But he drove off. They were shaken by the event, but nobody had fired a round. Unit NCOs went around to settle their soldiers down. Well, the driver came back and once again tried to run down the soldiers, hitting one. This time they fired nine rounds into the rear of that car. They knocked out both the tires and he still wouldn't stop. So they put one bullet in the window post and two in his head, and Victor Rivas was dead. Well, what did we find out? First of all, there were no license plates on the car. Number 2, there were rusted bullet holes in the car. This [being shot at] was nothing new to Victor, who was a Playboy gangster. The cops arrived on the scene very quickly. There weren't that many of them on the streets, but they were very responsive to calls for assistance. The first cops on the scene said "This is the SOB who tried to run us over about three weeks ago!" We thought that was rather interesting. The police found methamphetamines in the back of the automobile. That was the story of Victor, who was a convicted felon by the way. There was nothing innocent about Victor.

An interesting thing happened as fresh graffiti sprouted all over gangland. They didn't use this word "screw." You know the word they did use. The phrase "Don't screw with the Ninja Turtles" was spray painted around south central LA. You can see why they called soldiers the Ninja Turtles; you remember they [Ninja Turtles] were big in those days. Our kids probably don't know what they are anymore. But over there on the right side is a typical soldier with his helmet, face mask, flak vest and what all, and you can see why they were calling us the Ninja Turtles.

This is the other side of what we consider our other more serious incident.

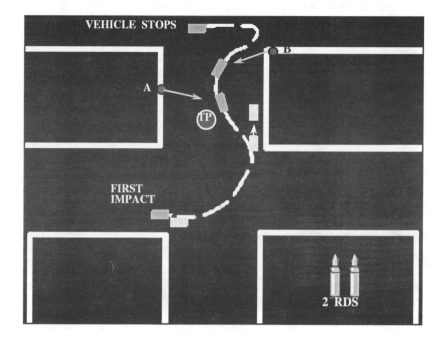

It occurred in Santa Monica on Santa Monica Boulevard. A blue Nissan Pathfinder rear-ended a white Volvo. The police pulled the drivers over here on to St. Andrews Place, a side street. A cop went over to take what they call a preliminary report. The guy in the Nissan was acting very strangely, gunning his engine and what all. He ran over the cop [who was later carried off to Cedars Sinai Hospital with a broken leg]. Then he tried hitting the soldiers, and they each fired one round. One got him in the butt, and the other made him a soprano. The perpetrator tried to walk off, but he was in no condition to walk. Now, what did we find out about this guy? He was on felony probation, 15 years probation from Florida for felony vehicular manslaughter. He was used to trying to kill people with his car. He had a .0179 blood alcohol level, much too high for California. He also had drugs on him. So these were the two people who were struck by our bullets. By the way, those kind of things are tough on soldiers, just as they are in combat, particularly when you're not shooting at a known, uniformed enemy. Our soldiers who fired needed psychological counseling afterwards. You need to be prepared for that, and we were. They had medical support immediately, people who worked with these young soldiers who had struck people.

Twenty-two total rounds. That's the total number of rounds we fired. Unlike in 1965, no innocent people got shot. The only soldier, and this is mentioned in the book [*Fires and Furies*, written by MG Delk], but I normally don't brief it, the only soldier who was struck by a bullet was a young PFC clearing his .45 who shot himself in the leg. That's more common than you might think. That's why every police station has what they call a clearing barrel. Those kinds of things happen.

What did we learn? I'm going to go through some general observations first.

I've personally been involved in four riots and other incidents. In every case, politics play a big role. Egos play a big role. Even some of us generals have egos, if you can believe that. And that kind of thing needs to be considered. It does play a role. Well, we happened to be in the biennial silly season during the riots of 1992. It was just before the primaries, and politicians were there in great numbers. Every time we had a gathering for purposes of creating peace or what all, they turned into political rallies as the politicians would show up with their signs. We got used to that. But you just need to know that went on. In addition, it is smart to have somebody who knows the local politics. Lawyers normally know local politics. The JAGs in the reserve components are lawyers, and so they gave us a lot of good counsel. You needed to know that Sheriff Block and Chief Gates don't talk and haven't talked for a year. And the mayor isn't talking to the chief of police. You need to know all of these kinds of things to avoid bear traps.

THE

E. O. C.

... WHERE UNCOMFORTABLE
OFFICIALS MEET IN
UNACCUSTOMED
SURROUNDINGS TO PLAY
UNFAMILIAR ROLES MAKING
UNPOPULAR DECISIONS
BASED ON INADEQUATE
INFORMATION IN MUCH TOO
LITTLE TIME

Emergency Operations Centers. Most of you GIs have seen that before. We sometimes call those Crisis Action Centers. They had the LAPD Emergency Operations Center all but shut down a couple of times as the politicians would come in there to have their discussions. What we needed was a super EOC ... with impressive maps, some public affairs types, and other people who had a handle on things so you could brief people in a venue not disruptive to operations. Politicians could make their policy decisions there, while soldiers and cops would get on about their operational business elsewhere. We did this after the riots were over, which was much too late.

Public affairs. Our public affairs were a disaster. Some of that was my fault. The trouble with poor public affairs is that it creates the environment in which we operate and it can get soldiers killed. I never did get a handle on the rumor spread by media types that our soldiers had no bullets. There was also the famous incident of the governor telling us to play Barney Fife: give one bullet to every soldier. Well, that's B.S.! No commander is going to do that with his troops, nor did we. The result was that gangbangers thought our soldiers didn't have bullets or they may have only had one. The result is that people think they can screw around with soldiers like Victor Rivas did. I didn't have a good handle on public affairs. You need an 800-pound gorilla running that show for you.

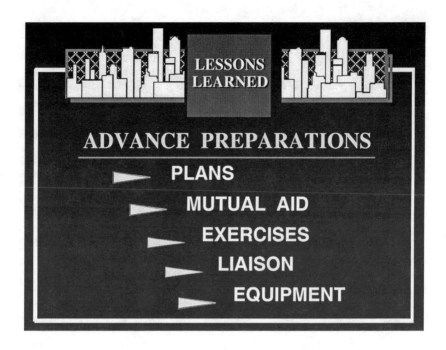

Advanced preparation plans, mutual aid, exercises. You need exercises, liaison officers, and equipment. That's all old hat to you guys.

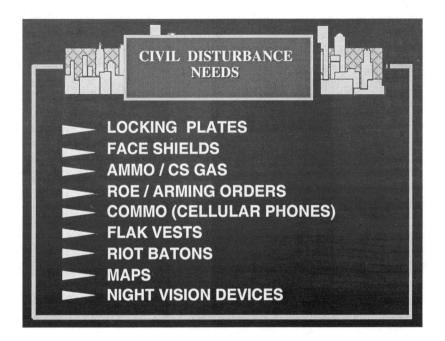

CIVIL DISTURBANCE NEEDS

► LOCKING PLATES
► FACE SHIELDS
► AMMO / CS GAS
► ROE / ARMING ORDERS
► COMMO (CELLULAR PHONES)
► FLAK VESTS
► RIOT BATONS
► MAPS
► NIGHT VISION DEVICES

What are some equipment needs? Body armor, face shields, ammo, and CS gas. We didn't use CS gas—the only time we used a lot of CS gas was in Berkeley and we did a great job of gassing a hospital. The trouble with gas is it doesn't always go where you want it to go, and you need to be aware of that. We've always been very aware of that and try to avoid using CS gas. It doesn't do any good to use CS gas on gangbangers anyhow. You use that on students or people that refuse to obey and can't shoot at you. I'll go into detail on Rules of Engagement and Arming Orders later.

Communications. Communications were lousy. We used cellular phones. Someday the military is going to have a radio that works well in an urban environment. They sure don't now. Even worse, you can't talk to the cops because your frequencies don't mesh (net). Flak vests you need, riot batons, no. You need riot batons in some situations; against students they will work. Something that works even better is bayonets. The riot batons are big, long suckers and they got in the way.

Maps. I want to talk about maps for just a second. In California, cops use Thomas Brothers. Do you have Thomas Brothers in the east? Well, whatever they use, you need to know what it is because if you go in there with a Corps of Engineers map you will find they are absolutely useless. You need to use the same thing the cops are using so you can communicate off the same map. So we have thousands of Thomas Brothers maps in our supply rooms. And night-vision devices . . . the military owns the night. There are a few night-vision devices out there in gangland, but we owned the night.

LESSONS LEARNED

RULES OF ENGAGEMENT/ ARMING ORDERS

Regarding Rules of Engagement and Arming Orders. I wouldn't change a word of those Rules of Engagement. They worked for us; they worked all the way through. A bright young major who used to be a cop designed them. Now, regarding Arming Orders.

ARMING ORDERS

	RIFLE	BAYONET	CHAMBER
AO - 1	SLING	SCABBARD	EMPTY
AO - 2	PORT	SCABBARD	EMPTY
AO - 3	SLING	FIXED	EMPTY
AO - 4	PORT	FIXED	EMPTY
AO - 5	PORT	FIXED	EMPTY / MAG IN
AO - 6	PORT	FIXED	LOCKED / LOADED

These are the Arming Orders. We were generally at AO-5, but without bayonets fixed. Fixed bayonets can be as much danger to your own soldiers as to the bad guys, so I recommend that you don't use bayonets unless you really need them. As I say, they're good against students, but gangbangers don't get that close to you anyhow. So we quickly modified that and AO-5 became without bayonets.

Communications and language. This is very important. A platoon's a different thing to a soldier. It's a different thing to a marine. It's a different thing to a tanker, and it's certainly a different thing to a cop. A platoon to many cops is 60 soldiers, while there may be only 20 or less in a tank platoon. But there are other communications differences. The worst incident occurred in Compton. In Compton, which was marine territory, two Compton police officers took a squad of marines with them and headed out to a domestic dispute. The cops walked up to the door, knocked, and the next thing you know someone fired bird shot through the door. One policeman was hit, but not hurt. His partner grabbed him and as he pulled him back he hollered to the marines "Cover me!" Now to a cop, that was very simple command. That means aim your rifle and use it if necessary. To a marine, and there were some well-trained young patriots in that squad, it meant something entirely different. They instantly opened up. A mom, a dad, and three children occupied that house. I later asked the Compton police department to count the bullet holes for me because there was a rumor going around there were 50 or so rounds fired. The police told me there were over 200 bullet holes. In some cases you couldn't tell how many bullets had gone through. They didn't hit anyone, but the point is, those great young marines did exactly what they're trained to do, but not what the police thought they requested. You need to understand the differences in language.

Training.

Useless. We wasted many hours on riot control formations. Now that works great when you've got a bayonet and you're coming at students. You try that with gangbangers and you're going to get your soldiers killed. What is needed is good, hard-nosed GI training. But what is especially needed is simply well disciplined young soldiers being good soldiers.

Intelligence. Very different. For the military, as you know, intelligence is predictive. Our intelligence officers fascinated the cops. We spent a lot of time working with them. Theirs is very event-oriented which doesn't work for us.

Transportation for a heavy division depends mostly on tracks. We couldn't use them; we didn't want to drive into Los Angeles with tanks and armored personnel carriers. We had to contract for buses, which meant that the civilians had to help us a lot.

Deployment strategies. This was controversial.

We sent our soldiers out two-by-two to take the streets back from the gangbangers. There's no such thing as a beat cop in Los Angeles. It's too dangerous for them to walk beats. We had to take the streets back from the gangbangers, and the only way to blanket South Central was to put soldiers out two-by-two. They ended up with two-by-two at every corner in some areas. One group could always could see other soldiers; they knew they could quickly get help if it was needed.

An E-5, a young sergeant E-5 fire team leader with a total of five soldiers had total responsibility for the Gateway Plaza Shopping Center.
My point is that we had to trust these young E-5s to do exactly what
you would hope they would do without a commissioned officer
standing there with his arm around them. And they did. He
[Sergeant David McGill] ran the show at that shopping center. This
young fellow is looking through binoculars at gang central right next
door. They were constantly being taunted and tested by gangbangers. Cars would come into that parking lot with the lights out
just to scare them. Rocks and bottles were thrown.

Staff Sergeant Brown is at the top of the burned-out service station looking catty-corner at the only un-burned-out business in the area, a 7-11. What he's looking at is all the gang and drug activity going on there. They didn't burn it out because it was their source for drugs.

That's an air policeman. I apologize for the quality of the photo, but I wanted to point out that the air police were working with us. We had them augmented with a chemical company and they handled protection of public utilities. We gave them that important mission and they did a great job.

Graffiti, two-by-two.

This became important: physical conditioning. The soldiers were on twelve-hour shifts, sleeping on the floor in markets and stores in sleeping bags. Physical conditioning became very important if they were to still make good decisions when they were tired. And they had every reason to be tired. They continued to make good decisions because we'd been pushing physical conditioning for years.

Exploit the differences. There is a difference.

Guardsmen seen differently than cops

There's a lot of enmity between the cops and the people they police, and so people wanted their Guardsmen there. The only exception to that was Salvadorans. The Guardia Nacional in El Salvador were the bad guys, and the minute our Guardsmen would grab a Salvadoran (we knew we had one when a Spanish-speaker would become extremely agitated), we quickly called for a cop. Then the Salvadoran would relax. They were scared to death of us.

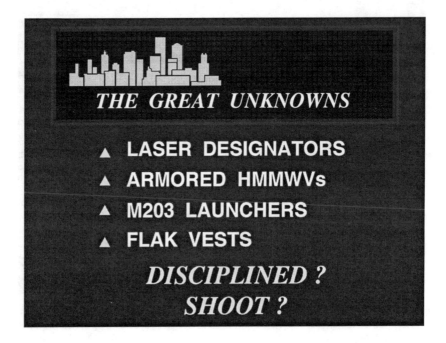

THE GREAT UNKNOWNS

▲ LASER DESIGNATORS

▲ ARMORED HMMWVs

▲ M203 LAUNCHERS

▲ FLAK VESTS

DISCIPLINED ?

SHOOT ?

The great unknowns. Laser designators. These things I'm holding in my hand [a laser pointer] were great for breaking up a crowd. If you shined it on a gangbanger's ankles and took it right up to the middle of his chest, it'd break up a crowd faster than anything. That worked great in those days. I'm not so sure it would now. As you know, every kid has one of these now. Interestingly, some of them also thought that our red flashlights were laser designators, which of course they weren't. Armored Humvees we talked about, and they were scared to death of M203 grenade launchers. We didn't have any ammunition for them, nor would we have used it if we did. But they didn't know that. Flak vests: they are so big as compared to the cop's, gangbangers were convinced that ours were better, more bulletproof. Of course, they aren't, but they didn't know that, so that worked for us.

SHOOTINGS			
DIRECTED AT THE MILITARY			
DATE	NATIONAL GUARD	MARINE CORPS	U.S. ARMY
30 APRIL	24		
1 MAY	10		
2 MAY	4		
3 MAY	4		
4 MAY	3	3	
5 MAY	11	1	
6 MAY	2		1
7 MAY	1		
8 MAY	1		
9 MAY	1		
10 MAY	2		
11 MAY	1		
12 MAY	1		

Shootings directed at the military. This is in addition to the bullets flying around any night regardless of the time of year. As I say, the riots ended on the second day. Cinco de Mayo [MG Delk pointed out another peak in firearms discharges corresponding to the fifth of May], that's what's happening there.

	DEAD	INJURED
CASUALTIES		
CIVILIANS	53	2,302
LAW ENFORCEMENT	0	71
FIREMEN	0	10
MILITARY	0	0
18,807 ARRESTS (2,628 FELONIES)		
5,383 FIRES		

Casualties: 53 dead civilians, 2,300 injured. You can see the number of injuries in law enforcement and firemen, but look at the military. I wasn't 'fessing up regarding our one soldier who shot himself in the leg.

It really was a miracle. It was enough to make a Christian out of a commander. When your soldiers are being shot at, you do a lot of praying. And I do consider it a miracle that none of our soldiers was hit. I would not count on that happening next time.

ANNEX 1: CIVIL DISTURBANCE MISSION TASKING GUIDELINES

		Federal	
Appropriate	State	Restore Law/Order	Preserve Law/Order
1. Main traffic control points	X	X	
2. Provide building security	X	X	X(–)
3. Escort emergency equipment	X	X	
4. Provide area security/area patrols	X	X	
5. Provide security at custody facilities	X	X	
6. Provide security for emergency work crews	X	X	X(–)
7. Protect sensitive sites	X	X	X(–)
8. Transportation for law enforcement personnel	X	X	
9. Show of force	X	X	
10. Disperse crowds	X	X	
11. Employ riot control agents	X	X	
12. Provide VIP protection/escort	X	X	
13. Provide reserve/quick reaction force	X	X	
14. Joint patrol/ride alongs	X	X	
15. Other missions mutually agreed upon	X	X	X
Inappropriate			
1. Hostage negotiation	X	X	X
2. Barricaded suspect	X	X	X
3. Evidentiary searches	X	X	X
4. Criminal investigation	X	X	X

ANNEX 2: RULES OF ENGAGEMENT
Los Angeles Riots, 29 April – 1 May 1992

I understand that I may be deployed to perform law enforcement support missions including crowd control, traffic control, perimeter security, protection of public safety employees such as firefighters, area security or roving patrols. I understand the following rules on the use of deadly and nondeadly force:

Nondeadly Force

1. Nondeadly force involves the use of physical contact, restraint, baton, M16A1/2 with bayonet or chemicals such as tear gas or MACE.

2. Nondeadly force will always be the minimum necessary to protect yourself, a team member, or a law enforcement officer, or in emergency situations.

3. Nondeadly force should only be used at the discretion of a superior officer or noncommissioned officer, a law enforcement officer, or in emergency situations.

Use of Deadly Force

1. Deadly force refers to the use of any type of physical force in a manner that could be reasonably expected to result in death whether or not death is the intent.

2. The use of deadly force is authorized only where all three of the following circumstances are present:

   a. All other means have been exhausted or are not readily available.

   b. The risk of death or serious bodily harm to innocent persons is not significantly increased by its use.

   c. The purpose of its use is one or more of the following:

      (1) Self-defense to avoid death or serious bodily harm (threat of harm is not restricted to firearms, but may include assault with bricks, pipes, or other heavy missiles, incendiary and explosive devices, or any other material which could cause death or serious bodily harm.

(2) Prevention of a crime which involves a substantial risk of death or serious bodily harm.

(3) Defense of others where there is a substantial risk of death or serious bodily harm.

(4) Detention or prevention of the escape of persons against whom the use of deadly force is authorized in subparagraphs (1), (2), and (3) above.

Civil Disturbance Training

I acknowledge that I have received basic civil disturbance training prior to my actual deployment in support of law enforcement.

I HAVE READ AND UNDERSTAND THE ABOVE USE OF DEADLY FORCE.

_____

Signed/Date

ARMING ORDERS

Arming Order	Rifle	Bayonet	Pistol	Baton	Magazine/ Chamber	Control*
AO-1	Sling	Scabbard	Holstered	Belt	In pouch/ empty	OIC/NCOIC
AO-2	Port	Scabbard	Holstered	Belt	In pouch/ empty	OIC/NCOIC
AO-3	Sling	Fixed	Holstered	Hand	In pouch/ empty	OIC/NCOIC
AO-4	Port	Fixed	Holstered	Hand	In pouch/ empty	OIC/NCOIC
AO-5	Port	Fixed	Holstered	Hand	In weapon/ empty	OIC/NCOIC
AO-6	Port	Fixed	In hand	Belt	In weapon/ locked	OIC

*OIC = officer in charge, NCOIC = noncommissioned officer in charge.

NOTE: In high-threat areas, soldiers were normally at a modified AO-5, _without_ bayonets fixed. This is much safer, and Arming Orders have since been modified to reflect this.

A MOUT DOCTRINAL CONCEPT
MG Robert H. Scales, Jr., USA

I appreciate the opportunity to be here. You're going to see something really extraordinary this afternoon: a serving two-star general in the American Army attempt to say something cogent without the use of PowerPoint.

I've been asked to come and talk about an article I wrote last October on urban warfare. In fact, I've just published a new article in *Strategic Review* in February [1999]. I commend it to you. Its called "Adaptive Enemies." It was written after a trip to the Far East during which I visited a very large Asian country. I had some incredible discussions about the American way of war and foreign perceptions regarding the American way of war with my counterparts there. Let me encapsulate them for you before I get to urban warfare.

Actually, the two articles are related. The "Adaptive Enemies" article says, essentially, that the United States has been displaying a firepower-centered style of war for the last 50 or 60 years. The premise of this thesis is that militaries are essentially complex adaptive organizations that tend, when faced with a threat to one style of war, to adapt themselves to another style. Most of the article is a recounting of the American experience beginning with the Pacific campaign in World War II where we faced essentially a thinking enemy with a will to win and time on his hands to adapt. For example, let me start with the terrible tragedy of Saipan (from the Japanese perspective), followed by General Ushijima Mitsurv's ability to reconstitute the Japanese 32nd Army in Okinawa. Ninety days of combat, 90,000 casualties, 9,000 dead Americans, and a Japanese defense that achieved its objective. We met an enemy who learned how to adapt.

In Korea, after the horrible experience of Lin Piao and the 1951 disaster, just north of Wonju, the Chinese adapted and came up with a system that allowed them to disperse and achieve their operational, and later their strategic, objectives. The Tet Offensive, a terrible experience for the North Vietnamese, changed their operational method. We suffered more casualties after Tet as the NVA [North Vietnamese Army] learned how to use a concept they borrowed from the Chinese: "maneuver under firepower." Other examples pervade. We could also talk about Afghanistan or the Israelis in Lebanon in 1982.

What I argue in the *Strategic Review* article is very simple: If an enemy is faced with an opponent who follows a predictable attrition style of war, that opponent essentially has an unbalanced military system. The enemy's object is to gain his operational objectives very quickly and then go to ground. He would disperse, then change his maneuver method from one of being objective-oriented to being area-oriented. This is an adaptation of the Maoist concept of area control. His object is to create a non-nodal entity, to distribute his area defense and build a military system that is able to prevail under a fire-power-intensive umbrella. His object is not to win but to avoid losing. He achieves this by using his own powers of time, the will to win, mass, and the inherent power of the defensive to hang on long enough to achieve his operational and perhaps his strategic objectives.

Last October, I also published an article about warfare in urban environments. I've spent thirty-some years reading and writing military history and looking at historical examples. Be careful when you use case studies, please. Remember the slide someone showed a little earlier that listed 15 or 20 examples of urban combat, and then they picked out a chosen few? Please be careful with that. If you develop operational doctrine from a selective use of history—this idea of niche history—you are walking right down the road to disaster. So please be careful. There's nothing wrong with a case study. But history is a holistic intellectual exercise. It's not a sort of library where you can go out and take one example from column A and two from column B to prove your point.

A couple other thoughts for you to consider. In our last Army After Next [AAN] War Game at the Army War College, we postulated an enemy whose objective was to quickly capture cities in the violated country in order to avoid American dominance in fire power. This game postulated war in the year 2025. The United States had a balanced military instrument, capable of achieving precision maneuver as well as precision fire, an instrument that was able to collapse an enemy very quickly, to strike, paralyze, exploit the paralytic effect of precision fires, control large areas, get to the battlefield before the enemy was able to set himself, collapse the opponent's will to win, and win quickly at minimum cost of life.

Well, unfortunately, that's not what happened, because the gentleman who played the National Command Authority waited, and waited, and waited. Finally, after the enemy had achieved his operational objectives, the AAN battle force and the AEF [Aerospace Expeditionary Force] companion force were dispatched late to the area of operations. Immediately the National Command Authority told his subordinates to get on with it, to get into the cities, and root the aggressor out. I found that to be very disturbing because it follows a pattern that I started to hear as a mantra during the last two or three years, as everybody sort of picked up on this idea about urban warfare. There's a tendency, again, if we don't look at history holistically, to go immediately into the tactical, to focus on the toys without thinking through exactly what it is you're talking about. Let me give you some facts, and then we can take our discussion from there.

It's the old "if, then" paradox. If the world is urbanized, if 50 percent of the world will be living in cities by the year 2025, then wars will likely occur in urban areas. That seems to be the reigning logic. That assumption is not historically grounded. In fact, if we look at the last 500 years of Western warfare, the number of urban warfare incidents has decreased, not increased. There has been far less urban warfare in the last 100 years then in the last 1,000 years. And there's a reason for it: Cities are strategic centers of gravity. They're valuable to someone. As Clausewitz says, in order for armies to join in battle, both sides have to agree to fight. Most incidents of modern urban warfare were mistakes, fought for all the wrong reasons: ideology, bad military doctrine, a political fixation, or an attempt to draw a superior power into city streets in order to bleed him. In other words, to snatch defeat from the jaws of victory. It doesn't matter what example we use, whether it's Stalingrad, Mogadishu, or Chechnya. When cities become political symbols or become lures that we can't avoid for some reason, they become battlegrounds. Look at the urban battlegrounds that we've fought on during the last few years. Every one of them was a political symbol. Hue, Manila, or Seoul in 1950. Why? First, we had an imperfect view of the nature of war. Second, the city became this lure, this magnet, if you will, that drew armies into it. Read *The Jewish War* by Flavius Josephus. It was written about 96 A.D. It is about the destruction of the second temple. If you read Josephus and read accounts of fighting in Chechnya, Stalingrad, Manila, or Seoul, you'll see many of the same psychological imperatives and political drives that caused enormous mistakes to be made. The truth is, in most major wars; cities are bypassed unless they are political centers or ideological lures.

Future enemies are unlikely to retreat into urban terrain. If an army retreats into urban terrain, he generally violates his own intent. The army gives up flexibility and the initiative. Don't let this enthusiasm to kick in doors and blow up buildings deflect us from the fact that urban terrain is terrain. When we prepare for war, we need to look at war first from the strategic perspective. What are our ends, ways, and means? Secondly, the operational aspect: What are the means we will employ to achieve our strategic objectives? And finally, what are the tactics we will need to employ in order to achieve those ends?

How many of you have read Liddel Hart's book on the indirect approach? He outlines an absolute immutable tenet of maneuver warfare: Always approach your objectives using the least predictable route. Yet suddenly, when we think about urban warfare, the whole idea of applying the indirect approach as a means of winning with minimal cost is somehow forgotten. During our war game, those who played the media and our political leaders continually asked: "What's taking so long? Why don't you reduce the cities?" That sort of created this feeding frenzy to get troops into the city and end it quickly. It appeared to me as curious that we're a country that seeks to win with minimum cost, not to win quickly, and yet here we were, whipping ourselves up to a froth to get it over with and expending casualties to save time. That's not the American way of war. We were taking the direct rather than the indirect approach. We were trying to destroy the city block by block, street by street, in order to save it.

Here's the point I'm trying to emphasize. How do we fight an enemy force by intruding into his territory? Back to my earlier statement: we treat it as a strategic, not a tactical problem. The first thing we seek to do is to place a force on the ground, in the open if we can, interpose our forces between the enemy and his operational objectives. This is the essence of maneuver warfare and the indirect approach. Since we live in a technological environment where the defensive is the stronger form of warfare, then our object is to exploit our strategic mobility, perhaps even our operational mobility, to place ourselves into a position to induce the enemy to come to us where we can exploit the full capacity of our fire power from defensive positions. It's what Longstreet tried to do at Gettysburg. It's what Lee sought to do at Chancellorsville. The tenets of war don't change.

If open warfare doesn't succeed, then we should seek a strategic *coup de main*. This is the case particularly if the aggressor is occupying the city of an ally—the Kuwaitis in Kuwait City, for example. We should attempt a sharp, quick, psychologically debilitating attempt to capture the center of gravity in that city and collapse it very quickly. In most instances the odds of that occurring successfully are low, however. I can't think of a successful effort in the last 50 or 60 years. We tried a *coup de main* in the AAN War Game. It almost worked. Unfortunately, the enemy had been in the city too long and managed to tactically set himself.

Up until this point, our objective was to win quickly. Time was our enemy. But once the enemy sets himself in the city, the time equation reverses itself. The urban defender, without the initiative, loses this time advantage and the attacker, external to the city, now has time on his side. Of course the time advantage is contingent on a couple of issues. Number one, before we go into a city we must be certain that we have control of the countryside. Most large urban areas cannot survive for very long without sustenance from the surrounding area. This is particularly true in many parts of the emerging world. The second issue is psychological. The nation has to understand why the pace of conflict slowed. A quick urban victory costs lives, while a methodical approach permits victory with fewer casualties.

So the enemy's retreated into the cities, and we now occupy the countryside. What do we do then? In many cities in the emerging world, the provision of basic services is tenuous even in the best of times. Therefore, the most assailable part of these urban areas is probably centered around their vulnerable infrastructures. Densely packed urban masses are easily manipulated by an information warfare campaign. We must maintain psychological dominance over the city so that all the information the citizens are getting about the situation is coming from us, not the enemy. Then we should cause just enough destruction to systematically take the city down by surgically taking out those installations that keep the city functioning, thus allowing the city to gradually collapse on itself. A sort of martial jujitsu, if you will.

What's the center of gravity of a city? The center of gravity is the population! We must find a way to control the population, and we've got to do it without butchering them. If we control the surrounding area, then we should be able to build a sanctuary. Creating a series of suburban safe areas into which the citizens can retreat leaves only the ruling elite and the army inside the city. Remember, what's another characteristic of large cities in these transitional states? Not only are they hanging by tenterhooks with regard to their being able to provide basic services, but they also contain all of the essential elements of government: cultural, political, economic, military, and bureaucratic, which often is the ruling elite. So we already have an urban tinderbox juxtaposed next to the urban poor ripe to be exploited. Control the enemy's information, find the spots in which the enemy is vulnerable, and then create sanctuaries.

Now, in order to defeat an urban enemy, even if time is on our side, we have to gain control of the urban mass. How do we do that? Well we do it like we did it in Vietnam, many times successfully: through the use of a cordon operation. A cordon operation is exactly what the name implies. It is a loose way to control the activities of the city from outside the immediate confines of the dense urban mass. We have the technology at our command, thanks to the information age, that allows us to do that today if we choose to. Control the city from outside; establish a loose cordon; take your time; find the vulnerable points; and destroy them systematically in a discreet, surgical form of direct action. The idea of pouring divisions into an urban mass to me is not wise. And the idea of carpet bombing is equally inappropriate. Remember that the decisive ingredient in urban warfare is the people. If we empty a city of its population, then all that remains is a militarily irrelevant urban shell.

Once we've established the cordon, and gained control of the urban masses, we've won. The war is over. We own the countryside. The enemy has no initiative. We've contained the enemy in the city. We can only lose by becoming impatient again; we don't want to snatch defeat from the jaws of victory. It seems to me that those urban takedowns that have been successful in the past—the Siege of Paris in 1870 comes to mind—have been the types of operations in which the army first gains the initiative; second, controls the countryside; third, contains the urban environment; fourth, reverses the relative advantage of time; and finally, allows the city to collapse on itself.

Remember, war is a test of will, not a test of material supremacy. An enemy is beaten when he knows he's beaten. The urban environment fundamentally is no different from any other military environment. We must protect the enemy's people and break the will of the army to resist. Once we've done that, victory is ours.

Questions and Answers with MG Scales

Major Sumner: The clock is often ticking before forces are committed to an operation. How would you overcome this time-over-mission type mentality?

General Scales: Excellent question. First, pick your wars carefully. Secondly, when we chose to go to war, make sure we understand what our center of gravity is. And our center of gravity is what? Will. Generally, it is defined by our tolerance for casualties. As an aside, I think we may at times overstate our aversion to casualties, frankly. Before the Gulf War, one of the things that struck me was the speculation that if Saddam Hussein used weapons of mass destruction, we would be suffering casualties of somewhere around 30,000. Not all dead, but casualties of around 30,000. America gasped, but we didn't stop the deployment. Nonetheless, our potential enemies view our weakness as will, as defined by our tolerance for casualties. So the answer is, we can have it slow, methodical with a low expenditure of casualties, or we can roar up the middle with lots of casualties.

Audience member: Would you say, then, that all this talk we've heard earlier today about operational objectives is incorrect? Are operational objectives immaterial?

General Scales: No sir. Operational objectives are important, but they are not necessarily strategic objectives. I can't think of an example in at least the last 100 years where a strategic objective could be defined as a single piece of urban terrain.

Audience member: Vicksburg?

General Scales: Okay, lets take a look at the real significance of Vicksburg. Fourth of July 1863. Lee is crushed. Twenty-four thousand casualties at Gettysburg. Pemberton surrenders to Grant. Forty thousand Confederate prisoners. By the fifth of July, was the South defeated? No, in fact the strategic center of gravity in the Civil War was the will of the Confederate nation to resist, not the ability to win or lose battles, or cities like Vicksburg for that matter. One could even argue that Vicksburg, since so little of the Southern sustenance came from west of the Mississippi, was not terribly relevant. Very little of what was needed to sustain the South came from west of the Mississippi.

Audience member: You talked about the center of gravity and that once all the conditions are met in the waiting game, time's on your side. Could you identify maybe a laundry list of tactical centers of gravity?

General Scales: Excellent question. No, let me stick to what I've said. The vulnerable part of a large urban mass is the civilian population. As a general rule, he who controls the civilian population will prevail. In taking down a city, our first object should be to surgically take out the ruling elite, if possible. That, however, is probably not possible, as we've seen in recent history. Absent that, then we need to take out those points within the city that will cause the urban mass to move out. What would that be? I'm not sure, but I could give you some examples. Certainly, I think communications could be first. Drop the communications and then establish our own surrogate communications network. Next? Probably those things that sustain life over time. Someone mentioned water. I'd say water is probably about right, electricity, and so forth. But in order to do that, and to do it in as humanitarian a way as we can, we need to give them an alternative way to survive. We need to have a sanctuary of some sort. Now, someone says, that's too much to ask for an army, to both provide succor for the enemy and then to defeat him. Not necessarily. We did it in Iraq fairly successfully. We leverage our occupation of the countryside to provide the sustenance for the civilian urban population as it seeks sanctuary. There is no urban mass that is economically viable or totally sustainable on its own. The city must get its sustenance from some external source. Power, water, and food. Well, we now control those, so we establish an alternative living environment outside the city to draw the civilian population out. If the enemy lets his population go, then he can no longer hide behind them. If the enemy keeps them contained inside the city, then he becomes the bad guy in the eyes of his own people because he can no longer sustain them. I'm giving you a very simplistic solution. Obviously war has too many variables, and it's far too complex to allow this model to be that clean, but the alternative is to go through this street-clearing operation and lose a lot of lives in the process. Be careful when your object drives you to immediately go to the tactical.

Audience member: To me, your solutions don't address the challenges that we actually face. We had to establish order in Haiti and in Mogadishu. Those are the challenges, I think.

General Scales: You are correct, but that's a different challenge. I mean, the area I'm discussing involves combat decisions. Haiti is obviously something entirely different. That's a humanitarian operation. But I believe that the principles of offering sustainment and sustenance to the local population are about the same. I'm talking about a city that's occupied by an army no longer able to sustain itself in the field. That army uses the city as a means of preserving itself long enough for us to tire of the war first. If you want to offer an alternative view, sure. I mean, Mogadishu's a good case in point; Bangladesh after the floods in 1991 would be another example. Different situation. One of the great wonders of the American military system is our ability to cover the full spectrum of operations. That's the subject of another discussion.

Audience member: General, I know you like a good argument. Let's say that we didn't stop the war after 100 hours. Say that we've got forces outside of Basra and Saddam Hussein is slaughtering his own population. Alternatively, what if Milosevic puts Albanians in selected urban areas in Kosovo and starts slaughtering them? We can't politically afford to stand there and let time take care of things. The enemy leader is not going to let the population leave the city. In these particular cases it appears we have a strategic, operational, and tactical necessity to go in.

General Scales: Give the National Command Authorities the option to make those decisions, but we have to be realistic. The Basra solution: Saddam Hussein goes in and starts executing his own citizens. But let me offer an alternative. Unless the enemy is a complete strategic incompetent, he is fighting a war to win it, or at least to avoid losing it. So he has his own strategic objectives. If his reaction is simply to line up people in front of CNN cameras and execute them en masse, he loses the support and good will of the rest of the world. That doesn't make a lot of strategic sense.

Audience member: What about what occurred in cities in northern Iraq?

General Scales: Again, give the National Command Authorities a realistic assessment of what's involved here. Let me give you a better example. Last spring we did a war game at the Army War College where a large Middle Eastern oil-producing state was invaded by another central Asian state. The invader went immediately into the cities. They didn't even stop to fight in the open but went immediately into the cities and started executing the extended royal family. The enemy used that as a means of inducing us to attack. And we did attack. And of course we got waxed. We suffered thousands of casualties in the game. To my mind, if we still view our center of gravity as being able to win at minimum cost in a reasonable time, then the National Command Authorities would be very reluctant to take that bait. But I'd also argue that the aggressor would be reluctant to offer it, because remember in this CNN world of ours, the enemy is also playing to world opinion as well.

Audience member: Let's go from the hypothetical to the actual. You've mentioned Tet and Hue. If we had tried the indirect approach and cordon and allowed the locals to come on out, I don't know how effective that would have been, given that the North Vietnamese slaughtered some 4,000 of the citizenry that were friendly to the South. They probably would not have let the locals come out.

Our enemies have a better track record of controlling the media than we do and even if they shut down the media for a while, they still have an advantage with the media that we sometimes don't enjoy.

General Scales: I'm not so sure that's true. I would argue that if we take information warfare as seriously as we do today, and we continue to exploit our advantage in that regard, and if our motives are as selfless as they have been in previous wars, I don't think that the information war will be lost. I think that Hue in 1968 was a terrible shock to the American people. And it wasn't just Hue. The whole country was in flames. And so the pressure was, as I recall, to take the enemy down everywhere as quickly as we could and expel him from several cities, to include Saigon. Did the extreme sacrifice of those Marines allow fewer South Vietnamese to die during that period of time? The answer may never be known.

Audience member: If someone is less optimistic about the relative advantage of time, is there an intermediate approach value besides the indirect approach?

General Scales: That's a great question. Let me answer it in a roundabout way. It seems to me that the Army is essentially a doctrine-based and doctrine-driven institution. We tend to do what we have either practiced or studied in the past. How long did it take the Civil War armies to adopt open order and the automatic use of field fortifications to offset the dominance of the muzzle-loading rifle? Four years, 600,000 dead. And that's because neither Army had time to think through what the impact of the extension of the killing zone might be. I would contend that the same rule applies in urban warfare. If we don't adapt our doctrine and change our culture to fit the circumstances, the soldiers, at least initially, are going to do what they've been taught to do at MOUT sites. Is there a middle ground? Perhaps so. But maybe we ought to say in our doctrine that in urban warfare, in most cases when fighting in cities, the time equation reverses. It's time to rethink maneuver warfare in a slightly different context. The next time we come up against the Citadel in Hue maybe we will take a moment and think about it.

Is there a middle ground for time, to get back to the original question? I don't know. I'll just tell you this: As a rule, armies that have been the most successful in urban warfare have been those who have not rushed the walls. They've been those who have been methodical about how they approach the city. Somewhere in our doctrine we have to be able to say that cities are different from open terrain. The principles of war still apply. Use the indirect approach. But the indirect approach in this case is not determined so much by availability of axes of advance at vulnerable points, but by the advantage of time.

Audience member: We're in the business of trying to figure out which toys are appropriate and affordable in conjunction with urban operations. I can't help but think that the concept of surrounding the enemy and taking him down indirectly seems to have application in the Kosovo situation right now. Milosevic is surrounded. He has perhaps one ally who's probably not going to come to his aid (Russia). He faces not an ad hoc coalition, but a real alliance, and what we seem to be doing is trying to make the cordon with an air campaign. There is the clarion call to make something happen on the ground.

General Scales: I'm having a tough time disagreeing with you. What's the hurry? My point exactly. It doesn't matter how we apply the pressure. Apply it from the air; apply from the ground; apply it from both. What's the hurry? Where's he going to go?

Audience member: I'm not sure I agree. We are painting ourselves into a box in Kosovo. We said we were never going to employ ground troops. We keep hearing about CNN diplomacy forcing us to do something. Look at all the refugees.

General Scales: What I say in my article is that the secret to success in future war is balance. I've got a new article coming out, in fact, and it's a companion piece to the "Adaptive Enemies" article. It's called "A Sword with Two Edges." In that article, I argue for the creation of a balanced military that has the ability to maneuver with the same precision in the future that we can strike with today. Let's say that the enemy adapts his doctrinal method of war to fight an attrition-centered style of war—ours. We can adapt, too, right? The gentleman I spoke to, a wonderful guy, retired Lieutenant General Lee from this large Asian nation I visited, used only one English term the whole time we talked. He used the phrase "non-nodal army." Interesting, isn't it? He said to me: "General, the information age is neutral. Mao was right; you were wrong." He noted that as we become more proficient at sensing, seeing, and striking targets with great precision through the use of the information age, his army will become better at his own style of war, which is to control large areas of territory with ever more discreet bits of combat forces (non-nodal armies) while still retaining the ability to mass on demand. We may have the advantage of satellite technology, but he has the advantage of cellular technology. It works to our advantage, just as much as yours does for you.

I found this gentleman to be relatively profound. If he's right, how do we develop a military force in the future that's able to take on an adaptive enemy who arrays himself to survive precision strikes? The answer for me is to build a military that's capable of precision maneuver. Our forces have to develop an operational doctrine or method of war that allows us to command territory over a broad expanse and yet still remain cohesive. To interpose our forces in ever more finite and discrete bits, to interpose ourselves between the enemy's forces, to break down his command and control, to prevent him from massing on demand, which is his center of gravity. Then we will seek to force the enemy to come to us, again presuming that the defensive is the superior form of warfare. If we do that, then this adaptive enemy I've just described to you is left with two unacceptable alternatives. Number one, he can stay in place and wither and die. Number two, he can come out of his holes and attack us, at which time the advantage is ours. Either way, he loses.

If you're developing a doctrine for a style of war that's already passed, you're simply making yesterday perfect. If we meet an adaptive enemy who's able to think a generation—or half a generation—ahead of you, then we're building a future force that's going down the wrong path. If we can't approach warfare with a balance of ability to gain positional advantage as well as to kill, then we are developing a military force that's in great danger of being defeated. We saw it in Korea; we saw it in Vietnam. How many more times do we have to repeat that?

Let me end with a quote. I wrote about this when I talk about the Western Front in World War I and how difficult it was for us to adapt to the dominance of the defensive. The war cost 11,000,000 dead before the generals could figure out how to do restore balance to the battlefield. Winston Churchill wrote, "Battles are won by slaughter and maneuver. The greater the general, the more he contributes in maneuver, the less he demands in slaughter." Balance, balance, balance is the key to future warfare. We may fight a major competitor in the future who not only has the ability to fight, to use a doctrine of area control, but also to apply his own form of precision and counterprecision. If we don't take him on with a force capable of both strike and maneuver, we run a real risk of slaughter and defeat. The danger is extreme.

Listen, I've enjoyed talking to you this afternoon. It was fun. Great conference and I appreciate the opportunity. Thanks.

NONLETHAL OPTIONS: FAILINGS AND FUTURES
Lieutenant Sid Heal, Los Angeles County
Sheriff's Department

Nonlethal Options
Failures and Futures

Sid Heal
Los Angles Sheriff's Department

I want to open by giving you a little bit of background. Most of what I'm going to focus on pertains to law enforcement, particularly non-lethal options for use in two major categories: mob riots and individual control. I would call this a grunt's perspective. I have no scientific degrees, but I have a lot of scars on my body from things that didn't work out just the way I had planned.

I had my first experience with a 1980 riot involving the Revolutionary Communist Worker's Party. That was my first exposure to what I would call a real, honest-to-goodness riot. The Revolutionary Communist Worker's Party crashed a black jazz concert during the Iranian hostage crisis. I can remember wading into mobs of people with all these placards—they were really substitute clubs; they'd nail them onto two-by-twos and they were beating us with them. Well, that was also my first lawsuit, which gave me a whole different perspective toward mob violence.

What I'd like to do is to provide one man's perspective. In fact, the subtitle of this presentation would probably be "The Emperor Has No Clothes." Like the little boy in the story, I'm just calling them like I see them.

An Idea Whose Time Has Come

- Useful for Interventions which Involve:
  - Urban Environments
  - Mixed Combatants and Civilians
  - Humanitarian Interventions - Peacekeeping, Peace Enforcement, Peace Making
- Strategic Implications
  - Lack of Justification to use Lethal Intervention provides impetus for Increased Provocation (Barking Dog Syndrome)

It has been estimated that more than 90 percent of law enforcement shootings occur within fifteen feet, take three to five seconds, and only involve two or three rounds fired. They are characterized by very close-in, short-range combat and by an almost instantaneous opportunity. I'll talk to you in a few minutes about snipers, because that's one of the other things we encounter.

Regarding peacekeeping and peace enforcement operations, what we're seeing right now is a blurring of war and crime. The military is rapidly becoming embroiled in peacekeeping and peacemaking, which is very similar in many respects to law enforcement, during which the mission never completely goes away. You never have this warm and fuzzy feeling that everything you did is right. And as we tell our people in the academy, when you're intervening between two combatants, there's 100 percent chance of making one of them mad.

On the other hand, law enforcement is also finding itself embroiled in military-like operations. During the 1992 riots, I was a watch and incident commander at Firestone Station. On the third day of the riots, I had nine 60-man platoons of deputies that worked for me, and I had two rifle companies from the National Guard and two rifle companies from the Marine Corps in an area of 4.4 square miles. It was unlike anything I'd ever experienced. It has some strategic implications. What happens is that the lack of justification to use lethal intervention provides an impetus for increased provocation, what we in law enforcement nicknamed the "barking dog" syndrome. Dogs are not intimidated by threats. A policeman can pull his gun at a dog, but because the dog doesn't know what the gun does, he has no idea what is going to happen. As a result, we're usually forced to shoot dogs. There is no intermediate option. We don't have an ability to intervene with nonlethal force.

Intervention Strategies

- Proper Employment is Key to Success
- Nonlethal Options
  - Are more Humane
  - Allow more Control
  - Provide More Flexibility and Freedom to Act
  - Are Less Likely to Provoke Others
  - Are Less Likely to Raise Public Outcry
- Military focuses on Intervention — Law Enforcement focuses on Prevention

There is no magic bullet yet. We don't have anything guaranteed to stop an individual. Lethal options have never been 100 percent effective. On the other hand, there's an unreal expectation that non-lethal devices will never kill. We've shot several people who didn't make it. We've accidentally killed people with nonlethal devices.

Nonlethal options are more humane. It's very difficult to make a case for a humanitarian operation if the only means we have of imposing our will is by killing the people we're sent to protect. The situations we become involved in tend to be evolutionary in nature. It takes substantial provocation to be able to justify our use of lethal force. Nonlethals allow more control because we're allowed to intervene at an earlier stage in the process.

The same thing applies with the military. With nonlethals, we can prevent escalation beyond the point where the only means of enforcing our will is with lethal force. It allows more flexibility and it's less likely to provoke others. One of the things we've discovered is that mobs are not homogeneous. The mob mentality that everybody describes exists, but not to the extent that everyone believes. There are different levels of commitment within a crowd. Many people are simply waiting to find out where the line is and how far they can push it. If we don't have the ability to intervene at an early enough stage, the line moves to a point beyond that we can live with. Then we're forced to take action.

Nonlethals are also less likely to raise public outcry. This is crucial, because we can win the battle and lose the war. In law enforcement, we're particularly sensitive to that, perhaps even hypersensitive, for the simple reason that we serve at the pleasure of the community. One of the things we tell people is that they don't get the amount of law enforcement they can afford, they get the amount they can tolerate. You can throw 100,000 policemen out there, and you can pay them and feed them, but if you don't let them do whatever they need to do, a lot of undesirable things are going to occur. That becomes crucial, because in some instances the public tolerance goes way down.

The effects of that intolerance can be long-lasting. Out of the 11,000 jobs that left L.A. after the 1992 riots, only 600 had returned two years later. Some people considered their insurance money as a gift. One of the people I talked to had been burned out in the 1965 riots and then had his new building insured for $350,000. His comment was that he could never have sold it in the middle of Watts for that amount of money. The insurance money was a gift. So the riot actually gave him the opportunity to get out of L.A. If you go back down to 102nd and Grape right now, the burned-out buildings are still there from the *1965* Watts riots! If you go down there right now, even the stores that have been rebuilt have no public entrances on the street side. There are no windows on the street side. They have essentially been built as fortresses around their parking lots.

One of the things I have noticed is that the military tends to focus on intervention, whereas law enforcement focuses on prevention. I would like to illustrate this point with a story. Two young boys decide to go on a walk in the woods. It's a beautiful day, and boys being what they are, they're having a good time. Then they encounter this bear. The bear starts chasing them and they start running. As time passes, they're getting tired and the bear's getting closer. They look back and it's obvious this bear's pretty hungry and its intentions are pretty obvious. Finally, one of the boys stops, takes off his day pack, and starts putting on running shoes.

His friend says, "What are you doing? You're *still* not going to be able to outrun the bear!"

His friend looks up and says, "I don't have to outrun the *bear*!"

One of these individuals saw the solution to his predicament as having to outrun the bear, while the other saw it as having to avoid being eaten by the bear. The one that had more precisely defined his situation had options available to him that the other one didn't have. Subtle differences in problem definition have a profound impact on problem resolution.

Nonlethal Weapons

- Defined by Intent — not Capability
- May not be Weapons at All!
  - May take the form of water, lights, or even smells
- Four classes of Nonlethal Weapons
  - Anti-Personnel (Most Well-Known)
  - Anti-Mobility
  - Area Delay or Denial
  - Anti-Infrastructure

For the first time in history we have a class of weapons that has been defined not by what it is capable of doing, but what the intended effects are. I'm a consultant for the U.S. Department of Justice on nonlethals; the only thing we could initially agree on is *not* to use the term "nonlethal." The term is virtually alien in the law enforcement world. When I was in Somalia, we had issues with foams, particularly sticky foam. I recall a conversation with our Staff Judge Advocate that went something like this:

"Gunner, what happens if you shoot somebody in the face?"

I said, "Sir?"

He says, "Will it stick their lips shut?"

I replied, "Yes, sir, it'll stick their lips shut."

He says, "Well, they'll die."

I said, "Yes, sir, and that's why *we* don't call it nonlethal."

There are four classes of nonlethal weapons. The first is anti-personnel. These are designed to stop an individual from doing something or to compel him to do something. Anti-mobility refers to being able to stop vehicles of some sort. Area delay or denial and anti-infrastructure are the last two. Anti-infrastructure devices attack entire systems such as communications, mass transit, and so forth. We have very little interest in anti-infrastructure in law enforcement. The one I'm going to spend the most time on is anti-personnel. In the Los Angeles Sheriff's Department our three big nonlethal areas are (1) intervention with nonlethal force, (2) stopping fleeing vehicles (which would be the equivalent of anti-mobility), and (3) detecting concealed weapons. These are the "bread and butter" of law enforcement functions.

Failures

- Nets
    - Single deployment, heavy, short-range, expensive, unable to separate individuals, requires clear background, easy to outmaneuver
- Foams
    - Messy, heavy, short-range, requires decontamination, severe logistical burden
- Both have limited applications and are Fiscal "Black Holes" — (No amount of Money will overcome Conceptual Flaws)

I need to warn you that I'm going to kill a few sacred cows here. I know that some of these projects have millions of dollars invested in them, but it's time to recognize that we've got about everything we're likely to get.

We've been messing around with nets since the 1970s when PCP was the drug of choice and we were desperate to find less lethal alternatives. The problem with nets is that they are conceptually flawed. You can only use them once. They are heavy and short-range. They don't work well inside buildings. You can't discriminate between people if they are standing next to each other. They don't work well if the guy is near vegetation like shrubs or trees. They're not very quick and a lot of times the target can outrun you. In addition, whoever employs the net is unarmed. He doesn't have the ability to defend himself. And it just goes on and on.

Another area of limited promise is foams. There may be some applications for foams, especially the rigid foams being developed by the Marine Corps Warfighting Lab, but against people they're messy, heavy, short range, require decontamination, and they incur a severe logistical burden.

Well, my point is this: these things are fiscal black holes. It doesn't matter how much money you throw at them. You can change the color, size, and shape. You can do whatever you want, but when you are done, it's still a net and it still has all the conceptual problems of a net. It's the same thing with the foams.

In the L.A. Sheriff's Department we spend an average $6.5 million every year just defending ourselves. We just lost a lawsuit in the spring of 1997. It was the largest payout in the history of law enforcement: $23 million on a single lawsuit. So for the first time in history, it's cost-effective to go out and find effective nonlethal weapons. The other thing is, if we employ a nonlethal, and it fails and we end up killing the guy, we've sent an implicit message: we tried not to use lethal force, but the technology was just not there.

Futures of Nonlethal

- Potential for Near Term Success
- Strategic — Most promising is by Focusing on Equipment
  - Employment is Less Likely to be Delayed by Bio-Effects Concerns
  - Few Humanitarian Issues
- Tactical — Nerve Stimulation
  - Technically Difficult but Potentially Most Rewarding

There is potential for near-term success. The most promising area appears to be equipment. This is because there are almost no bio-effects to be concerned with. One of the things that continually causes problems is bio-effects, the human biological part of all this. Sometimes, even when we have a device that we know works, it will take a year to get it certified.

A tactical area that we're really interested in right now is nerve stimulation. It's technically challenging but potentially the most rewarding. To give you an example, if you hit your funny bone, your arm instantly goes numb. You don't have the ability to close your hand, it gets cold, and your entire attention, for a brief period of time, is focused on your body. When the pain goes away, there's virtually no harm; there's no bruise; there's nothing. If we had the ability to artificially recreate that effect, it would provide a tremendous tactical advantage.

The only thing we have that's even close right now is the Taser, and the only way we have to deliver it is with wires. As a result, we create a medical casualty because we actually drive these probes right into the guy. We're looking at a device that uses glue and appears to have potential. They call it rhinoceros snot because they make it in San Diego right next to the zoo. This stuff is about as sticky as stuff can get, and it actually conducts through clothing.

The Search for the "Holy Grail"

- ## The "Magic Bullet"
  - Will yield a Change in Behavior
  - Will Employ a "Holistic" Approach
    - Synergy may be Created by Combining Technologies
  - May take the form of a Threat, particularly an Implied Threat
    - Social, Political and Cultural Implications cannot be ignored
    - The Object is not to fight — The Object is to Win!

Finding the best nonlethal device is like the search for the Holy Grail. That's what a lot of these civil libertarian groups would have us find. I don't see it ending anytime in the near future.

Two examples very quickly: snipers. A sniper's center of gravity is the ability to remain anonymous, particularly in the city. But his critical vulnerability is that once he's detected, he has no sustainability. Sooner or later we will win. We're looking at two devices right now that will track snipers; one from the Naval Research Lab and one from a group called Trilon Technologies. If we have the ability to instantaneously locate and identify one of these guys, we may have the ability to greatly reduce the threat due to sniping. To give you an idea of the magnitude of the problem, consider the last study that I did right after the riots. I found that about 800 people were killed by gangs each year. Overall about 1,550 were murdered with guns. This is not all murders; this is just those involving guns. And about 4,000 people are shot each year that survive. I don't even know, to be honest with you; I'd have to actually sit down and count all the partners that I've had that have been shot, including some that have been killed.

Let's take a general, holistic look at nonlethals. People are comfort-seeking creatures. What happens if it's 12 degrees out there and you're using water? It becomes a nonlethal device all by itself. The other thing is, if it freezes onto the pavement, it becomes an anti-traction technology.

One of the things that happened while we were in Mogadishu was that the Somalis somehow got an idea that the lasers we had would sterilize them. That's an important thing when you're in the United States and you want to have a family, but in a Muslim culture, that's an extremely serious thing. Now lasers do a lot of things, but sterilizing people is not one of them. We never said they'd sterilize anyone, but it was to our advantage not to say they wouldn't. It became a force multiplier all by itself.

Nonlethals may take the form of a threat, particularly an implied threat. Let's be honest. A credible threat of lethal force is a nonlethal option. We've been holding people at gunpoint for as long as we've had guns. And I can tell you right now that it's pretty language independent. If you have a situation in which you don't have the ability to do a lot of things, when the gun comes out, everything becomes focused very quickly. The advantage of the threat is that usually what we *can* do and what we're *willing* to do lie pretty far apart. The vast majority of the times I pull my gun, I have no intention of using it. Now that goes against the grain of a lot of the civil libertarian people, but what happens is that I create the perception that I could use it, and therefore the guy has to go through a thought process that basically becomes a force multiplier all by itself.

Developmental Issues

- Long Term Goals
  - Success is the Biggest Obstacle
    - Bias for Kinetic Energy Munitions
    - Good Enough becomes the Enemy of Better
- Fully Mature Nonlethal Force will probably be judged by a Combination of Two Criteria
  - Will Provide Adequate Defense against Lethal Force
  - Crooks will use it along with, or instead of, Lethal Force

Now to long-term goals. We have a bias for kinetic energy weapons. Basically the things that we're doing, we're still doing with kinetic energy. We're just not hitting them hard enough to cause the injuries that we'd cause with lead. The good-enough becomes the enemy of better because if it's working at all, it's better than what we had before, and there's really not a lot of impetus to go out and find something better yet. Success, in a sense, has been our biggest obstacle to further advances.

So the fully mature nonlethal force is probably going to be judged by two criteria: (1) will it provide adequate defense against lethal force? Right now I can tell you that it does not. When we currently use nonlethal options, we always have a guy assigned to protect our nonlethal officer. And he has a lethal weapon. The guy with the fire extinguisher, the pepper spray, the gas, or whatever else doesn't have the ability to defend himself. (2) The other judge of nonlethal effectiveness will be whether the crooks use it. Pepper spray is great stuff. It's been the best thing that's happened to us for a long time. It's the most useful nonlethal device we've had. It's effective about 85 percent of the time. Now if you do a robbery in California, you're going to get a five-year enhancement if you use a gun, without the possibility of probation or parole. So you have a two-year sentence for a robbery and five more years for gun use. That's a seven-year minimum. You could do the same thing with pepper spray and get only the two years. But the preferred weapons for robbery are still guns and knives. Why? Because they are so much more effective.

Public Expectations

Phaser

Star Trek

September 1966

- Criteria
  - Highly Portable
  - Discriminating
  - Instantaneous Effects
  - 100% Effective
  - 100% Reversible
  - 100% Safe
  - Environmentally Benign

This is really the public expectation. I did some research on this because this really is the public expectation. This first appeared on September 8, 1966, a Thursday night, some 33 years ago. Everything I've read estimates that any individual with access to a television in the last decade has some knowledge of this device. If I had just put the picture up or just used the sound, probably everyone in this room would have been able to identify it, describe it, and list everything that it does. This is the ideal. It has become the public expectation. The key point to remember is that this has *never existed*. But it has set the benchmark that we're expected to live up to. It's highly portable. It's discriminating. It's got instantaneous effects. It's 100 percent effective, 100 percent reversible, 100 percent safe, and environmentally benign. I was just on a radio program with Daryl Gates from the LAPD. He's got his own radio program now, and one of the comments that he made during the interview was that he'd always thought that one of the government organizations had one of these programs on the black side; law enforcement calls it the dark side; they were just holding it out because of some strategic reason to keep it secret. And I said, "Well sir, that's very possible. I don't have access to everything, but do you really think the federal government has both the ability to do that and keep it a secret for all these years?"

Nonlethal Developmental Strategy

- None
  - All current Nonlethal Weapons and Programs are Tactical in Nature
  - Bias for Kinetic Munitions
    - Pain is OK — Trauma is Not!
- Training is all but nonexistent in Military
  - Nonlethal Employment Methods are Thoroughly Integrated into Law Enforcement training

Now, regarding a nonlethal strategy. First of all, there is none, at least on the law enforcement side, and as best as I can see, the same is true from the military side. All current nonlethal weapons and programs are tactical in nature, at least all the ones I've seen. As I've said, we still have this bias for kinetic energy. I gave a presentation at Johns Hopkins University and the Institute for Nonlethal Defense technologies at Penn State. One of the things I said was that we don't need the phaser. I mean, we'd like to get it, but recognizing that we're not going to get it, anything that you can do to cause the pain without the trauma is okay with us. Pain is OK, it's the trauma we're trying to avoid.

Training on nonlethals is all but nonexistent in the military. That's starting to change, but it hasn't changed to the extent that it has in law enforcement. In law enforcement, it's inculcated into everything that we do from the very beginning. That's critical because right now the key to success has been the application: when to do it and how long to do it, how much to apply and under what circumstances.

> # The Line Between Science and Fiction
>
> - Go for the "Low Hanging Fruit"
>   - −Minimize Expense and Effort — Maximize Effectiveness
> - The Standard is not Perfection — the Standard is the Alternative
>   - −Optimal Time Lapse is < 3/4 of a second
>   - −Optimal Distance for an Individual is >21 feet
>   - −Optimal Distance for a Mob is >40 yards

Go for the low-hanging fruit, the technologies that offer the most promise with the least amount of effort. Those that don't require a commitment to climb into the tree but are close enough to take now. The standard is *not* perfection. The standard is the alternative. We realize we're not going to get the phaser; what we're asking for is anything better than what we already have. And in some cases, that's not a lot to ask, because most of what we've got as far as nonlethals boils down to pepper spray and a baton. If you look at the nonlethal devices that are on the market today, you can boil them down into two broad categories: kinetic energy or chemical weapons. There's not a lot of directed energy. There's not a lot of other alternatives. There are three considerations regarding standards that I believe are fundamental. One, the optimal time lapse between the time that we fire a weapon and the time it becomes effective has to be less than 3/4 of a second. That is about how long it takes for a human to process a stimulus and respond. It is why we have tissue damage when we touch a hot object before our brain can tell us to move our hand. If we don't have the ability to impact an individual in 3/4 of a second, he can respond. As a result, we have to have a counter. In fact, to the degree that we can't predict what his response is, we have to have more than one. The situation is complicated logarithmically.

The second thing: the optimal distance for application has to be greater than 21 feet. Twenty-one feet is crucial because that is the limit from which most people have the ability to close and kill you with an edged weapon. And that's why you will hear of recent incidents such as those in New York, Riverside County, and Florida where an excess of 40-some shots were fired at people with knives. What happens is this: even if you have the gun drawn, a guy can attack you and he can still kill you with the knife if he's within 21 feet. That's about 7 yards.

In law enforcement our longest range we qualify at is generally about 25 yards. The vast majority of our firearms engagements are at 7 or 8 yards—the length of a car or the inside of a room. In the Marine Corps, our minimum qualification range is 200 yards!

Finally, the optimal distance for a mob is greater than 40 yards, because they can usually throw things at you and hit you with rocks and bottles at that range. We call that the stand-off distance, the distance between our formations and the mob.

I don't know if you've noticed this or not, but in most cities that haven't been shot all up, there are not a lot of rocks. There are not a lot of bottles. So if you have a riot planned, you have to bring your own stuff. The Nazis are great for this. They bring golf balls and spark plugs. I bring this up because they can throw those about 60 yards. Anything better than 40 yards we're willing to live with, but 60 yards would really be our goal. So we need weapons that have the ability to impact people farther out than that.

Questions?

Lt. Sid Heal

Los Angeles Sheriff's Department
Special Projects
1275 N. Eastern Ave.
Los Angeles, CA 90063

323-980-2202
CSHeal@lasd.org

Audience member: I'm from the MP school. We're working on a system to halt vehicles. There was a test done a couple weeks ago during which we stopped an 18-wheeler going about 35 miles an hour. I think that's pretty successful. Are you looking at that?

Heal: Yes, sir. We think it has great potential because it also locks the doors; the guy can't get out.

Same audience member: It seems from the tone of your presentation that you're kind of skeptical about nonlethal weapons. At the MP school, we are very excited about them. We're also in the preventative business. We think nonlethals include use of UAVs. Maybe a UAV could fly over the battlefield and shut down vehicle engines. We also have a machine gun grenade launcher. We're testing it now with a nonlethal munition that will take a target down at up to 60 yards. I think that'll be pretty effective.

Heal: Thank you, sir. I hope I didn't leave the impression with you that I was skeptical. I tend to be a pragmatist; I'll vouch for that. But no, we are also pretty excited about some of these things that are happening. I guess if I had to summarize it, it's not skepticism but impatience.

Audience member: Based on your experience in Somalia and Haiti, do people misread our willingness to use force if we employ non-lethals?

Heal: I understand the implications of the question: are we sending a signal that we lack resolve or have a reticence to use force? I don't know if we're sending the signal, but some people are going to take it that way. That's why when you look at law enforcement we don't separate nonlethal force from lethal force. Part of the difficulty is that we can't predict the effectiveness of nonlethals as accurately as we would like. For example, let's say I'm going to employ a sponge grenade and someone asks me how will it work. I have to say, "Well that depends. What's he wearing? Where did you hit him? Was he under the influence of any drugs? How cold was it? Cold makes a big difference because some of these devices get harder when it gets cold. On the other hand, people wear heavier clothing when it's cold. As a result of that, there are a variety of factors that all have to be factored in. It doesn't lend itself very well to predicting how effective a weapon is going to be. One of the things we tell law enforcement is if it doesn't work the first time, repeat as necessary. So any device that shoots more than once is more appealing than a single-shot device.

Audience member: You were saying that directed energy is not an area that seems to have a lot of promise. We've heard things like acoustic weapons oft times have unintended physiological effects. Have you had any experience with acoustic or microwave systems?

Heal: Experience, no.

Same audience member: Actually my question was about acoustic weapons as a directed energy device. We've heard that there are some physiological side effects.

Heal: That's one of the systems that has captured the attention of the press. I must have said fifty times that they're not advanced enough so that we're going to be able to use them, at least not in the next five years. We focus on 60 months or less for reasons that I won't go into right now. They potentially have great promise. It would be very exciting if they could make it work, but from every-thing I see they will not be ready in the next five years.

THE CITY'S MANY FACES
Gen Terrance R. Dake, USMC

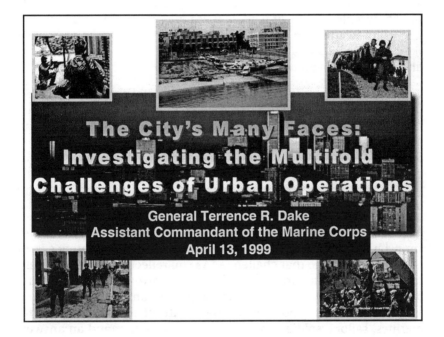

Editor's note: The text found on these slides approximates the remarks made by General Dake on April 13, 1999. Because the General does not use notes when speaking, his remarks were not a verbatim recitation of these speaker notes.

Good evening. I would like to thank the sponsors of this conference for their efforts to put such an important and timely event together.

The issues you have brought this audience to address are critical if we are to achieve success on the battlefields of tomorrow. I also want to thank each of you who have taken the time to attend and for your contributions to our efforts to improve our ability to conduct urban operations. Alone we will not find the answers but together we can create solutions to the myriad challenges posed by urban operations.

As the new millennium approaches, society is undergoing a transformation about which many have written and lectured. We are in an age of uncertainty that has been labeled, among other things, the information age, a "clash of civilizations" (Prof. Samuel Huntington), and the "urban century" (Marine Corps Intelligence Activity). We do not know which, if any or if all, of these titles is accurate. But we can rest assured that the armed forces will be called upon to pursue national interests in a chaotic world against a variety of opponents and most likely in an urban area.

The challenge for us is to determine how societal changes will transform the conduct of warfare. We need to determine if our existing military doctrine, force structure, and equipment will be effective in a new security environment. Each of us should be asking a very basic question—will societal shifts and demographic changes make our current military operationally irrelevant, and if so, can we take steps now to ensure we are prepared to meet the nation's needs in future crises and conflicts?

A specific challenge we face is how to project power into the urban sprawl in which much of humanity will reside. This conference is an attempt to examine that challenge. As you reflect on today's presentations and listen to those scheduled for tomorrow, keep in mind my basic question. Our nation demands that we answer this question, and many others prior to our next battlefield challenge. The marines, sailors, soldiers, and airmen you lead demand an answer. This is no small task, so I encourage you to devote every effort to this undertaking.

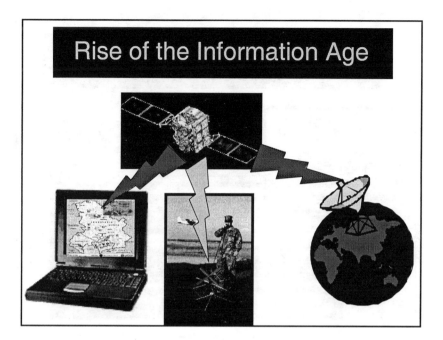

How will societal changes impact military operations in general and urban operations specifically? The question gets at the heart of this conference. War is and will remain a clash of human wills. If the society in which those wills are created, developed, and influenced is altered, there is a great propensity for war to also be altered. While the nature of war—that is, its inherent violence, chaos, and danger—will remain constant, we do not know the full extent to which societal changes will alter the character and conduct of war. I venture to say that it will be immense.

One of the most significant changes is the proliferation of information technology. Societies are being radically altered by the impact of the personal computer, the Internet and the World Wide Web, cellular communications systems, and the like. The transition from industrial societies to information societies is dramatic. Digital technology is shrinking the impact of geography on global events to a minimal consideration while simultaneously increasing the amount of information available to individuals. We must strive to understand the impact on military operations.

The ideological shackles of the Cold War system are gone. No longer do two superpowers seek to control violence in order to prevent a world war. Nations, city states, ethnic groups, and religious societies are all feeling the freedom to pursue self-determination without the restraint imposed by the United States and the former Soviet Union. According to the Carnegie Commission on Preventing Deadly Conflict, "a significant source of conflict is to be found in the competition to fill power vacuums, especially during times of transition within the states and often as a result of the end of the Cold War and the collapse of the Soviet Union."

The Carnegie Commission further specified that "much of the violence wracking the world since 1989 has been attributed initially to ethnic causes, rooted in immutable history, or to the unavoidable release of tension or redress of grievances held too long in check by the last vestiges of colonialism or the bipolar international structure. The conflicts in the former Yugoslavia and Rwanda, for example, have had ethnic, historical, and broad social components, but they also have had a strong, immediate political component. In these cases, the precipitating motivation for conflict stemmed from actions designed to achieve political goals. Leaders within factions steered the conflict toward violence, tapping into long-standing, deep-rooted ethnic tension as an accelerator. Later, those historical and ethnic forces surged out of control, fed by momentum, suffering, and acts of retribution."

The rise of intrastate conflict and the relative decrease in interstate conflict during this decade is an indication that political battles are occurring within states as people and groups vie for power. Many of the leaders involved in these essentially domestic political battles are choosing military means to achieve victory.

Demographics

I would like to present some statistical data to demonstrate the dramatic upheaval the world is experiencing. This information was compiled variously from the United Nations, the Carnegie Commission on Preventing Deadly Conflict, and the Marine Corps Intelligence Activity.

- Over the past 40 years, the number of urban dwellers has more than tripled, from 737 million to 2.5 billion.

- By 2005, the majority of the world's population will live in urban areas.

- Over 160,000 people per day migrate to developing world cities.

- New York City reached 8 million in 150 years, while Mexico City and Sao Paulo will reach 8 million in 25 years.

- By 2000, 21 of the world's largest 30 cities and 264 of its 414 "million" cities will be in the developing world.

- The UN predicts that while the population of the developed world will increase by 50 percent by 2050, the developing world population will rise 80 percent in the same period. Significantly, one-fifth of the world's population accounts for less than 2 percent of global income.

While developed nations are able to plan for increases in population and to develop the infrastructure necessary to support large concentrations of humanity, the majority of nations and states in the developing world are incapable of similar achievements. They are literally being overrun by humanity. Thus the developing world is more prone to conflict as the lack of resources and services causes competition to reach violent levels. We must understand how these demographics will compound the impact of the societal shifts previously mentioned.

Editor's note: The graphic comes from a NOAA Web site. "The Nighttime Lights of the World data set contains the first satellite-based global inventory of human settlements, derived from nighttime data from the Defense Meteorological Satellite Program (DMSP) Operational Linescan System (OLS)."]

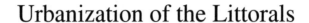

Urbanization of the Littorals

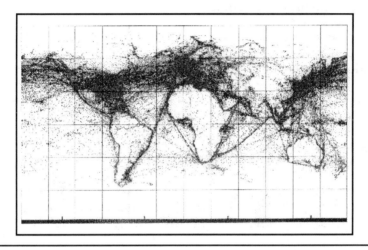

Recent UN forecasts predict that 85 percent of the world's population will reside in urbanized areas by the year 2025. We are witnessing a massive worldwide migration from agrarian to industrialized societies. What you see on the screen is a representation of human activity across the face of the globe. The location of humanity along the littoral regions of the world is indisputable.

Please note that the main image on this slide does not contain a map. I say again . . . there is no map on this slide. The image you see was produced by depicting the data collected by sensors. What you see is a summary of electronic activity as measured over a 30-day period. If there are any doubts about the importance of the littorals, sea lines of communications, or urban areas, then this image should dispel them.

You may conclude for yourself what areas of the world are important for mankind. You may also conclude what locations will most likely serve as the battlefields of tomorrow. As the global trend toward urbanization increases, the military significance of cities is likely to increase proportionally.

Why Operate in the City?

- **Panama 1989**
- **Liberia 1990 & 1991**
- **Manila 1990**
- **Kuwait City 1991**
- **Mogadishu 1991**
- **Haiti 1991 & 1993-4**
- **Los Angeles 1992**
- **Mogadishu 1992-94**
- **Bujumbara 1994**
- **Kobe 1995**

- **Mogadishu 1995**
- **Monrovia 1996**
- **Atlanta 1996**
- **Tuzla 1996**
- **Zagreb 1996**
- **Tirana 1997**
- **Kinshasa 1997**
- **Freetown 1997**
- **Mombassa 1998**
- **Asmara 1998**
- **Pristina 1999 (?)**

There are many within the defense establishment who do not believe we should allocate significant resources toward the development of capabilities necessary for urban operations. I politely but strongly disagree. To those who espouse the idea that we can bypass or lay siege to a modern city, I ask them to reconsider their position. To those who believe that there are no valid reasons to pursue military operations, I ask them to reconsider their thinking. To those who state that combat in cities is too difficult, I challenge them to contribute to our efforts. There are myriad reasons that we may have to fight in the city.

Over the last ten years, American military personnel have operated in the urban areas depicted on the screen before you. If you extend this timeframe back another seven years, we can include Beirut, Grenada, and Tripoli. I would like to emphasize that this list does not include the major wars of this century and the many urban operations from Bastogne to Manila, and from Inchon-Seoul to Hue City.

The list includes humanitarian assistance, security, noncombatant evacuation, support to domestic authority, and combat operations. These operations represent the day-in and day-out reality for our forward-deployed forces. Our marines, sailors, airmen, and soldiers are living, fighting, and dying in urban areas around the world because that is where many of America's national interests reside. When those interests are put at risk, we must be prepared to take the necessary action, which may include urban operations, to further national objectives.

In addition to the strategic and operational requirements for conducting urban operations, I believe we are moving out of an agricultural and industrial era where the majority of the world's population lived in rural or suburban areas to an information age wherein most humans will reside in an urban setting. This dramatic shift of the world's population will only serve to increase the need to create military capabilities adequate to successfully execute urban operations at an acceptable cost. I would like to present the factors that lead me to believe we are entering an urban century.

Maneuver Warfare in the City

"The worst policy is to attack cities. Attack cities only when there is no alternative."

Sun Tzu

"It's the only place they [our future adversaries] can take our technology and mute it...We avoid the cities, but that is where we will be taken-so we had better learn to fight and win in the city."

General C. C. Krulak

At first glance it seems that these two military leaders are espousing opposite, even contradictory, points of view. Sun Tzu warns us of the dangers of fighting in the city. The wise strategist, according to Sun Tzu, "skilled in war [will] subdue the enemy's army without battle" and will "capture his cities without assaulting them." We should take seriously the warning Sun Tzu provides of the dangers of fighting in the city.

General Krulak has told us that the enemy "watched CNN" and learned that we will defeat him if he engages us in open battle. Therefore the logical move is for our future adversary to select close terrain wherein he can attempt to meet us on an even battlefield. We may have no alternative to fighting an enemy in the city. We can attempt to besiege the city, but that move risks creating an enclave from which and in which he can operate and survive. We can attempt to bypass him and ignore his moves against us, but that risks providing him with targets to attack. It also requires us to divert much-needed combat power toward protecting lines of communication. We can attempt to blast him out, but that risks destroying the very city we are attempting to save. We can attempt to attack him directly, but that risks suffering heavy casualties. Therefore, if our opponents choose to give battle in the city and refuse to come into the open, we are in a quandary as to how to engage them.

Keeping with our warfighting philosophy of maneuver warfare, we must maintain the ability to pick the time, location, and manner of engaging an opponent. This is the challenge we face regardless of whom or where we fight. This challenge is greatly complicated when the opponent elects to operate in an urban area. We must develop the means to engage and defeat an enemy who selects the city as his battleground at the time and place of our choosing. How do we accomplish this task while addressing the risks outlined above?

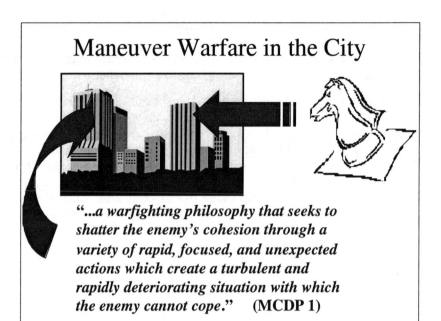

Maneuver Warfare in the City

"...a warfighting philosophy that seeks to shatter the enemy's cohesion through a variety of rapid, focused, and unexpected actions which create a turbulent and rapidly deteriorating situation with which the enemy cannot cope."   **(MCDP 1)**

The aim of Sun Tzu's strategy was to mentally defeat the opposing general and thus achieve a physical battlefield victory. *"His primary target is the mind of the opposing commander . . . "* The maneuver warfare philosophy of the Marine Corps holds that same objective as the top priority.

We are now faced with the challenge of applying this philosophy to a situation where the opponent has purposely selected a battlefield that "mutes" our traditional strengths of mobility and firepower. His selection of the urban battlefield gives him an asymmetric advantage. Therefore *we must examine how to develop new means of employing firepower on and providing mobility to a battlefield that traditionally hinders the use of these warfighting capabilities.*

As mentioned briefly, our history includes many urban operations. The idea of conducting urban operations is nothing new and is certainly no revelation. However, we believe that we can and must execute urban operations in new ways.

We seek to find ways to avoid destroying the city in order to save it. In fact we . . .

Operational Challenge

Defeat the adversary...

...while maintaining
the viability of the
urban infrastructure...

...in order to
ensure survival
of noncombatants.

There will soon be six billion human beings on Earth. According to the latest population estimates released by the United Nations, this will happen on October 12, 1999. As noted previously, most of these people will rely on the services provided by the urban centers of various societies. Any approach to conducting urban operations must build on a foundation that considers the impact of two features, people and infrastructure, on military activities.

The military commander must approach MOUT with a philosophy that views the city or urban area as a living entity rather than as a battleground or just a piece of terrain. He must understand that the city is more than a battlefield; it is a home, a place of business, a source of nourishment, a seat of government, as well as a location of religious, cultural, and social significance.

The operational challenge for the joint force commander is to design and execute a campaign that will resemble a surgeon cutting away a cancerous growth while keeping the patient alive. This statement should not be construed as meaning that hard, violent fighting will not be needed. We will in fact need to maintain the capability to conduct traditional, tactical-level, conventional operations against a determined foe. However, the campaign design must ensure that all urban operations are not reduced to tactical fights consisting of block-clearing, house-to-house fighting. If we successfully restore mobility to the urban battlefield and create precise fires of minimal collateral damage, we should be able to selectively execute these operations at critical nodes within a city.

As always, the prudent military commander will focus on the enemy, but urban operations provide the added dimensions of protecting and providing for the survival of noncombatants and minimizing damage to the infrastructure upon which they depend for survival. We must not destroy the city in order to save it. We must also not allow the enemy to achieve an asymmetric advantage by choosing to fight in a city and stripping us of our firepower and mobility advantages.

Our attempts to conduct urban operations are complicated by the fact that the missions we will be called upon to accomplish span the spectrum of conflict. We must simultaneously be able to field a force capable of winning a major theater war that is fought in a region that includes urban centers while also preparing that same force to execute Operations Other Than War in similar geographic settings. Military personnel will be called upon to distribute food, execute police functions, fight a determined foe, and deal with the destruction of major portions of a city.

The two distinguishing features of cities, people and infrastructure, will have equal influence on military activities in major theater war and operations other than war. Our preparations for tomorrow's conflicts and crises must address the entire spectrum of conflict. We must not attempt to categorize urban operations as either a MTW or an OOTW scenario. Urban operations will conceivably be conducted at every point along the spectrum of conflict.

One fundamental question comes to mind: Can a military organization trained, equipped, and prepared to conduct a major urban battle also be capable of conducting humanitarian assistance operations in the city? The question applies equally in the opposite direction: Can a military force prepared to conduct a noncombatant evacuation from an embassy also be capable of warfare of the utmost intensity? These are the difficult questions I encourage you to wrestle with, debate, and try to solve. The Marine Corps believes the answer is a resounding "yes." The key to fielding this type of military organization rests squarely on the shoulders of our junior leaders. I commend to each of you an article that appeared in the January 1999 issue of the *Marine Corps Gazette*. In this article, our Commandant states: "In many cases, the individual Marine will be the most conspicuous symbol of American foreign policy and will potentially influence not only the immediate tactical situation, but the operational and strategic levels as well . . . he will become, as the title of this article suggests, the *Strategic Corporal*."

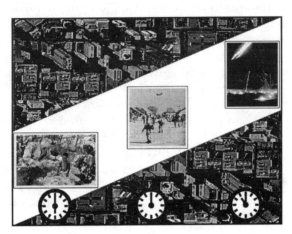

The Three Block War

Strategic - Operational - Tactical

The Commandant of the Marine Corps has described a scenario of the future that he calls the "three-block war." In simple terms we believe a marine will be engaged in humanitarian assistance at sunrise, peacekeeping at noon, and conventional combat at sundown. These three activities will all occur within a three-block area on the same day. What are the implications of such a scenario? While I encourage you to consider the impact on equipment design, weapon system requirements, force structure, command and control, intelligence, and logistics, I charge you to focus on recruiting, training, and education.

What are the requirements for adequately preparing our young service members for the challenge of the complex, high-stakes, asymmetrical three-block war? How do we develop junior leaders prepared to deal decisively with the sort of real-world challenges that they will confront? How do we instill the mental agility necessary to apply the philosophy of maneuver warfare during rapid-paced, extremely deadly, chaotic urban firefight?

One answer lies in recruiting and another in professional development. We must continue to recruit high-quality men and women of character. We must not lower our standards. The challenges a young NCO will face in the three-block war demand intelligence, creativity, resilience, and a strength of character. These traits and characteristics must be nurtured and developed through a demanding training and education curriculum. Such a curriculum must emphasize leadership, integrity, courage, initiative, decisiveness, mental agility, and personal accountability. We must provide our NCOs with the training and education that will enable them to successfully negotiate the obstacles of the three-block war.

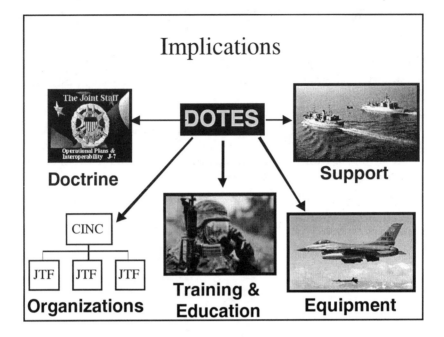

The implications of my views on urban operations cut across doctrine, organization, training and education, equipment, and support (DOTES). I believe there are no single-point solutions—no silver bullets. We must pursue solutions to the challenges we will confront in the urban century with a holistic approach. In the Marine Corps we utilize the Concept Based Requirements System to develop new operational capabilities. This tool allows us to examine the impact of a concept across all the categories of DOTES.

The characteristics we will need throughout our operational forces are:

- Flexibility
- Agility
- Multifunctionality

We must field people, organizations, and equipment that can execute multiple missions in a variety of situations throughout the spectrum of conflict. We need a military force that can alter its output based not upon a total refit of the organization but through mental agility. In short, the focus of our efforts to create more effective capabilities for urban operations is the individual marine, airman, soldier, or sailor.

The single most important implication of the urban century is how we recruit, train, and educate our personnel. We must, as previously mentioned, recruit men and women of character who have the potential to be conceptual problem solvers.

In addition to these implications I would like to cover two other topics before I close. The first is a need and the second is a first-order issue that we must resolve if we are to make significant progress toward improving our capabilities to conduct urban operations.

Mission or Environment?

There is significant discussion within the Joint community on the following related question: "Are urban operations a distinct 'mission area,' or do military operations on urbanized terrain represent an 'environment' in which existing doctrine and forces may effectively operate without significant alteration?" Stated differently: "Are military activities in urban areas so distinct that they require a different doctrinal approach than other mission areas, or does a city simply present another, more complex terrain set?" If the first part of these questions can be answered in the affirmative, than we must examine the entire spectrum of issues related to doctrine, organization, training and education, equipment, and support systems (DOTES). If it is the second part that better describes the situation, improvements in tactics, techniques, and procedures will suffice to help us prepare for future urban operations.

The Marine Corps has not determined the answer to these questions. We continue to examine urban operations to determine whether they represent more than a complex environment and if they merit treatment as a mission area.

Whether you agree with us or not on this issue is less important than the fact that we, as a defense establishment, address the challenges of conducting urban operations. What is not debatable is that urbanized terrain poses a complex and challenging set of circumstances in which to conduct military operations. It has all the characteristics of the natural landscape, coupled with manmade construction, resulting in an incredibly complicated and fluid environment that influences the conduct of military operations in unique ways. This manmade construction dramatically increases the complexity of warfare and creates a multitude of friction points with which a military force must deal. The addition of large numbers of people in a dense area, in extremely close proximity to combatants, further enhances the difficulties of conducting military operations on urbanized terrain. We must investigate the requirements for using military force in an urban operation and then determine how to meet those requirements.

The Future...

Joint Pub 3-06

- **J-8 Urban Working Group**
- **Doctrine**
- **Experimentation**

The Joint Staff

Force Structure, Resources & Assessment   J-8

**Marine
Corps
Warfighting
Laboratory**

In closing I would like to provide a glimpse into where are we headed. We have started down the path we need to be on. The creation of Joint Pub. 3-06, *Joint Doctrine for Urban Operations*, is a tremendous step toward a future warfighting capability. This particular document is notable in that it is the result of the JWCA process. Many of the attendees of this conference worked tirelessly in a joint effort to convince the services, the Joint Staff, and the CINCs that this document is needed. There are other positive signs, such as the

- Continuation of experimentation in urban operations, such as the Army–Marine Corps MOUT ACTD, the Army's Dismounted Battlespace Battle Lab, and the Marine Corps' Urban Warrior;

- Development of Minimum Collateral Damage Weapons for Urban CAS by the Air Force and the creation of the Urban CAS Range by the Marine Corps;

- Efforts by the Joint Special Operations Command to examine the need for a SOCOM urban operations concept;

- Insertion of language in OSD and JCS planning documents that addresses the reality of urban operations;

- Creation of Service Urban Working Groups or Task Forces in three of the four services.

These steps demonstrate that we have come a long way in just a few short years. But let no one here rest on their laurels, for these efforts only scratch the surface. We all have much more work ahead of us than behind us before we can truly say we have addressed the needs of urban operations. A fertile field has been plowed due to efforts of many people in this audience. We must take advantage and plant the seeds that will grow into a capable joint urban operations capability.

Questions

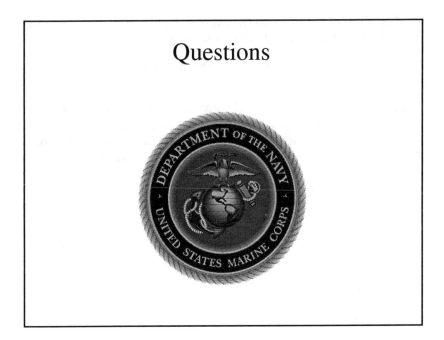

Before I conclude and we engage in a dialogue, on behalf of General Charles Krulak, the 31st Commandant of the Marine Corps, I extend a debt of gratitude to the organizers and sponsors of this conference. You have helped the Department of Defense begin to address the complexities we will face in the urban century.

I also would like to extend my personal thank you for the honor of serving as your keynote speaker. I hope that in some small way my remarks have done justice to your gracious invitation.

I look forward to your questions.

PREEMPTION FOR MOUT: REVISITING AN OLD STRATEGY
Dr. Randy Steeb, RAND

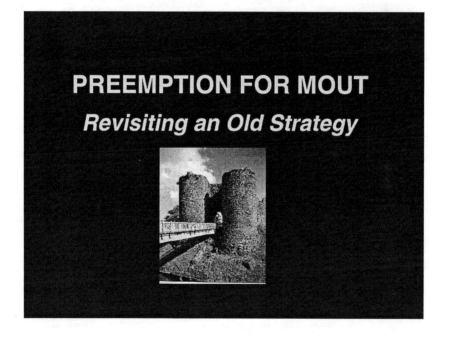

This annotated briefing describes new concepts of operation for urban warfare with an emphasis on defensive preparation rather than eviction of an enemy after he has become entrenched in a city or town. The work includes motivation for investigating the problem, description of alternative concepts, and discussion of synergies between the concepts and expected outcomes. This research is preliminary, with the intent of creating a dialogue on such concepts. The effort represents a portion of the work being performed on the Arroyo MOUT project at RAND under the sponsorship of SARDA.

Background

- Significant tactical combat advantage lies with defending force in MOUT environment

  - Recent history suggests anywhere from 8-10 to 1 advantage to defender

  - Small forces can hold off much larger forces, even when "siege" tactics have been employed

- Nonetheless, current emphasis of MOUT is on improving the *offensive*, attack operations

- While offense may be essential, it is prudent to plan to shift more of the battle to defense

History, exercises, and analysis have all shown that forcible eviction of an entrenched defender in an urban area is a daunting proposition, and heavy casualties may be expected. Many of the development thrusts of the MOUT ACTD and the focus of service exercises in urban areas have been on such offensive operations, however. It may be far more efficient and expedient, on the other hand, to invest in means to protect urban areas at risk: shifting the mission to delay, disruption, and defense rather than flushing out a dug-in defender.

Objective

- Explore improving MOUT through a more comprehensive preemptive strategy that includes:

  – Identifying early urban preparatory methods and corresponding technologies

  – Examining extremely responsive and reactive force capabilities

  – Considering the impact of indirect or "offset" concepts

There are three major components to the comprehensive preemptive strategy we envision for future MOUT operations. The first is preemptive preparation of the avenues of approach and interiors of an urban area at risk. This should slow the attacker and make him a better target for the second two components of the strategy—rapid reaction forces and "offset" engagement. Rapid reaction forces are ground, air, and naval forces quickly brought in to isolate and attrit the enemy force, while offset engagement concepts are used to disable and attrit any enemy forces that do penetrate the urban areas. In this way, the concept for operations embodies a mix of offensive and defensive components at both the tactical and operational levels.

The organization of the document is straightforward. First we look at motivations to investigate the problem by examining the options open to the United States in urban areas. Next, we look at the basic components of a preemptive strategy, along with recent enabling technologies. The third section describes how a typical engagement might occur, and the final section summarizes our observations.

Current MOUT Focus May Be Too Narrow

- Appears to be well below strategic level (e.g., tactical and operational)
  - ACTD focus on tactical
  - USMC Urban Warrior focus on operational
- May resolve near-term deficiencies but may not respond to evolving nature of threat (future, advanced threat)
- Does not account for possible manpower-to-threat-mismatch

The motivation for redirecting MOUT toward preemption springs from several sources. Key MOUT initiatives—the MOUT ACTD and the Urban Warrior experiment—tend to be necessarily limited in focus and extent because they have to demonstrate new capabilities in the very near term. There is still an emphasis on close combat with either a tactical or operational level of application. These efforts may make room-to-room and building-by-building clearance some-what faster and more survivable, but they do not solve the larger problems of dealing with evolving threats or widening conflicts. Each of these issues will be discussed in the following slides.

Visit with Engineer School Illustrated Difficulty of Breaching a Modern Urban Defense

Tiered defense concept*

- **Defense of city begins well outside city borders**
  - Obstacle belts, WAMs, AHMs
  - Fires from artillery and direct fire
- **Border is heavily defended**
  - MOPMS, Volcano, and other obstacles
- **Streets of city defended**
  - Claymores (command detonated)
  - Snipers
- **Interior of buildings defended**
  - Booby traps
  - Demolitions to drop building

} Infantry weapons superimposed

* Camp Lejeune MOUT site

For example, many MOUT demonstrations involve friendly forces quickly dislodging enemy infantry occupying a few buildings. To see what a capable opponent might do in this situation, we presented a plan map of Camp Lejeune to U.S. military officers at the Engineer School at Fort Leonard Wood, and asked them to prepare a defensive plan for this complex. They quickly developed a tiered defensive concept in which an engineer platoon, along with an infantry company for overwatching fires, could effectively defend the town against a battalion or possibly a brigade of attacking Blue forces. Approximately four truckloads of equipment were estimated to be needed—smart mines, command detonated explosives, booby traps, and building demolitions were all part of the notional plan. The idea is that the force would engage the attacking force at range along the likely avenues of approach, continue to activate belts of defenses as they approached, and then bring down the buildings themselves as the Blue force penetrated. It was noted that the Engineer School had previously used this type of defense very successfully in exercises at Fort Hood.

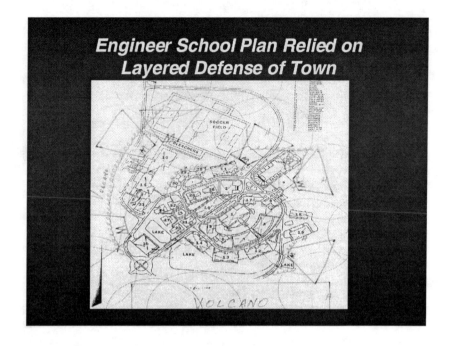

Engineer School Plan Relied on Layered Defense of Town

The Engineer School plan is diagrammed here over the layout of the Camp Lejeune MOUT site. They defended the town utilizing Blue systems and tactics. The outer layer of defenses are shown by circles representing wide area mines (H = Hornet, the wide area mine designation) and Volcano (helicopter-deployed anti-tank mines). Interior to these are MOPMS (Modular Pack Mine System) command detonated mines and concertina wire. Finally, all buildings were rigged with explosive charges and could have been demolished.

The plan shown is limited to Blue options. If carried out by enemy forces instead, it could have been expanded to include anti-personnel mines, booby traps, use of hostages as shields, and even application of weapons of mass destruction.

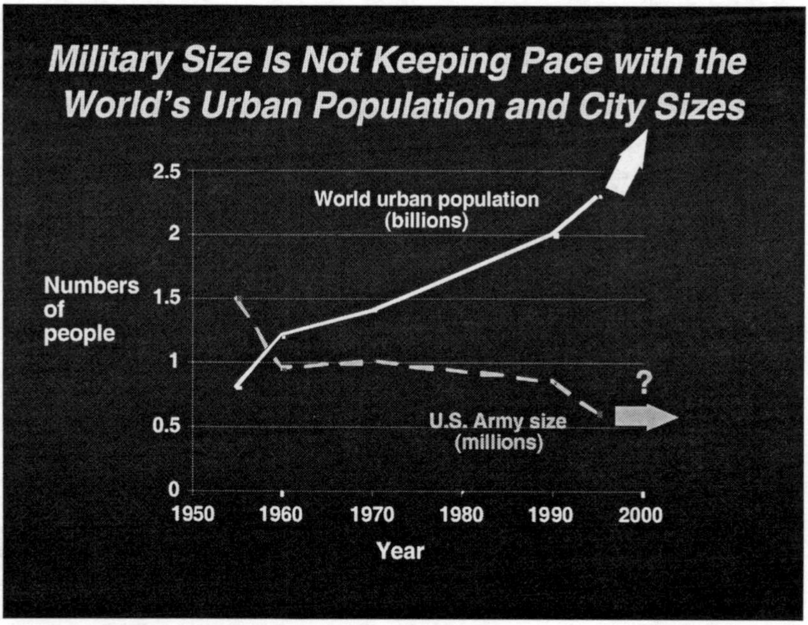

Military Size Is Not Keeping Pace with the World's Urban Population and City Sizes

Even if such an attack could be successfully carried out against a few buildings, the U.S. military would be hard pressed to free an urban area having tall buildings or large defended areas. Worldwide urban population trends and U.S. military strength, as shown in the above chart, highlight the increasing difficulty of the problem.

This section will delineate the basic components of the preemptive strategy.

A Preemptive MOUT Strategy Might Involve Three Critical Components

1) Indigenous preparation of urban environment

2) Highly responsive forces

3) Adopting new "offset" capability

These are the three critical components of the strategy. Each will be described in turn.

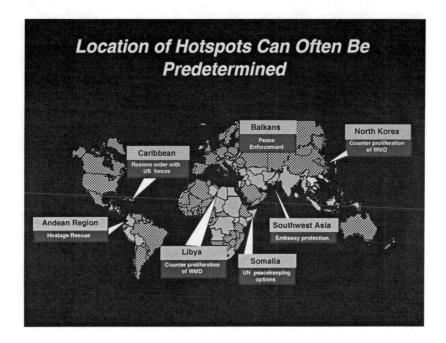

Preemptive preparation of urban areas begins with identification of the areas at risk. Potential hotspots around the globe are highlighted in this figure.

As noted earlier, a rapid reaction force may be equipped with sophisticated equipment, but may not be able to deploy to the urban area for several days.[1] In this time the enemy could have taken the city, forcing the rapid reaction force to attack an entrenched foe.

A more effective option might be to prepare our allies' vulnerable cities with many of the technologies described. These systems could then delay and attrit the advancing enemy, buying time for the rapid reaction force. Such an action might even deter an enemy from considering an attack on the urban area.

---

[1]Even the very light Division Ready Brigade of the 82nd Airborne, tasked to be the "first to fight", requires as much as 4-6 days to deploy to distant locations such as Southwest Asia.

Preparing the Cities May Constitute Most Effective Part of the Strategy*

- Prepare cities at risk by equipping and training the local forces, and by installing networks of systems

- Initiate defensive systems to delay, isolate, or attrit invading forces, buying time for rapid reaction forces

- Isolate and engage slowed enemy systems using fires from rapid reaction forces and standoff weapons

* The first two steps are similar to 1994 Chechen preparations of Grozny

The first steps in this preparation/preemption process are to instrument the city with secure, surreptitious sensors and to pass this information on to both decisionmakers locally and back to CONUS. The sensors should be able to collect information about enemy activities, locations, force sizes, battle damage, weapon types, noncombatant status, environmental conditions, and road trafficability. There should even be intrusion sensors in some key buildings which indicate occupancy.

Distribution of the resulting information needs to be made through redundant secure lines. These will probably need to be high bandwidth in order to support voice, data, and images and to overcome bandwidth loss due to encryption. Software radios will need to be given to trained, trusted members of the local populace to facilitate passage of HUMINT. These radios will also have to be controllable in case they fall into enemy hands.

Each Defensive Prep Function Requires Different Technologies (2 of 3)

- Ensure infrastructure protection
  - Design protected systems to deny enemy control of water, power, gas, phones, etc.
  - Proliferate backup power, comms
- Install obstacle network to slow or disable invasion
  - Emplace controllable vehicle barriers (barricades, superlubricants, foams, combustion inhibitors, RF bombs)
  - Emplace jammers with notch frequencies for friendly comms and navigation
  - Place smoke generators at critical locations

Those structures critical to the safety and well-being of the urban area will have to be protected. There will also need to be backup generators, communication systems, emergency water supplies, and medical treatment facilities.

A key element in the protection of the urban area is the installation of remotely controlled obstacle networks. Pop-up commercial vehicle barriers are currently available; these can be placed on avenues of approach to the built-up area (such as those currently in place in South Korea) or on city streets themselves. More futuristic obstacles, such as superlubricants, sticky foams, and smoke generators, can be located at key canalizing points. Jammers, RF bombs, and other canister or munitions-type systems can be placed or lofted as needed.

Each Defensive Prep Function Requires Different Technologies (3 of 3)

- Control and protect inhabitants
  - Ensure sufficient emergency/medical shelters, underground evacuation routes
  - Install public communications system
  - Distribute "tags" for noncombatants, friendlies
- Coordinate actions using mobile command centers
  - Acquire mobile command and control vehicles with planning and visualization tools, comms within area, and contact with rapid reaction forces
  - Load databases with structural maps, critical node analyses, population characteristics
- (Extreme case) Install remotely controlled lethal and nonlethal weapons

Noncombatant control can take the form of ensuring marked routes to shelters, dispensing tags for noncombatants and neutral forces, and even using calmatives or incapacitators for controlling mobs.

All of these actions need to be controlled using command centers. Some of these should be mobile to avoid detection, localization, and neutralization by the enemy, while others can be located in protected structures. The C2 centers should have access to building blueprints, city infrastructure plans, and other information needed to coordinate a response.

The most extreme case for preparation of the city is installation of weapons that can be remotely detonated. These, of course, have the greatest chance of being subverted by the enemy or activated accidentally. They should be implemented only in areas of high risk; they must be kept under close control.

A Preemptive MOUT Strategy Might Involve Three Critical Components

1) Indigenous preparation of urban environment

➡ 2) Highly responsive forces

3) Adopting new "offset" capability

The second component involves insertion of and coordination with a rapid reaction force. Unfortunately, this force may take many days to arrive.

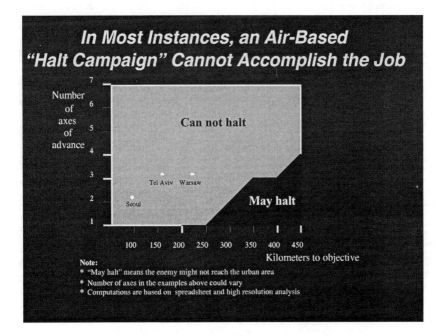

Even air force "halt" options, in which bombers are to deploy quickly and deliver hundreds or even thousands of smart munitions and GPS-guided weapons, require time and space to employ. The above chart, summarizing outcomes from high-resolution simulation runs and a weapon effects computational model, shows that many real-world situations are not suited to such responses.[2] The enemy has plenty of time to enter and secure key objectives in cities such as Seoul and Tel Aviv before bombers can significantly attrit them.

---

[2]Data analysis and chart provided by John Gordon IV, RAND.

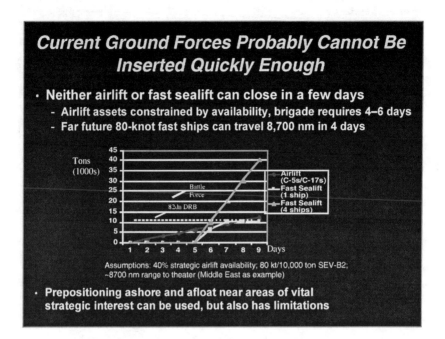

In a similar fashion, ground forces are also limited in their ability to close quickly and block an enemy from reaching key urban areas. The above figure shows that even a very light force (82nd Airborne) requires many days to close and be ready for combat. Heavier, more mechanized forces may be able to close quickly with future fast sealift, but even these may arrive too late to prevent an enemy from occupying key urban areas.

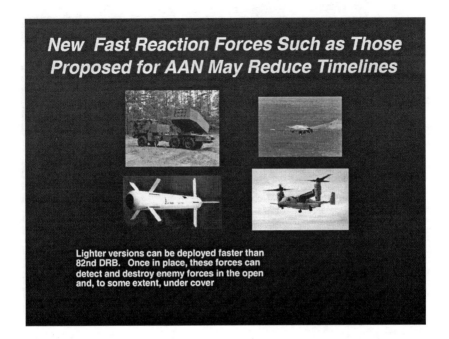

New Fast Reaction Forces Such as Those Proposed for AAN May Reduce Timelines

Lighter versions can be deployed faster than 82nd DRB. Once in place, these forces can detect and destroy enemy forces in the open and, to some extent, under cover

Many concepts have been proposed for light fast-reaction forces. These include more mechanized concepts such as the U.S. Army's Army After Next battle force, the USMC's Sea Dragon concept, and lighter, less-maneuverable force concepts such as DARPA's Small Unit Operations teams. These concepts potentially have extensive amounts of precision firepower available, but they will still require at least several days to close.[3]

---

[3]Images of V-22 courtesy of Boeing Aircraft Inc., fiber optic guide missile courtesy of Janes International.

A Preemptive MOUT Strategy Might Involve Three Critical Components

1) Indigenous preparation of urban environment

2) Highly responsive forces

➡ 3) Adopting new "offset" capability

The last component of the strategy is the use of new "offset" capabilities for engaging an enemy in urban areas from standoff locations.

Advanced Concepts for MOUT: Are Other Options Available to U.S. Forces?

- **Focus has been on close combat eviction of enemy**

- **However, our examination of advanced concepts notes:**
  - Many aspects of intelligence gathering, planning, and maneuver can be carried out at standoff or from behind cover
  - New technologies may allow physical and organizational isolation of enemy units
  - Precision fires can add surgical attrition of exterior and some interior targets

Much of the training and preparation for MOUT has emphasized close combat. In fact, the focus of the MOUT ACTD appears to be on near-term demonstration of new technologies primarily designed to aid the warfighter in room-to-room, building-by-building clearing.[4]

In this section we will discuss options that should minimize this high-attrition close combat through the use of remote means for intelligence gathering, isolation of enemy units in the city, and precision fires from range.

---

[4]For a detailed discussion of the MOUT ACTD and of current doctrine and training status, see the ACTD web site (http://yorktown.dc.isx.com/mout/) and Glenn (1998).

New "Offset" Concepts May Be Able to Force Combat to Outside of Buildings and Cities

Key concepts:
- Sector-and-seal
- Nodal warfare
- Noncombatant control

• Underlying principles
- Localize combat to key areas, nodes
- Sector off key areas
- Force combat to outside of buildings
- Turn defenders into unfavorable locations
- Interior engagements use robotics

Three concepts make up our "offset" approach to MOUT. Two of these, termed "sector and seal" and "nodal warfare," represent major departures from current tactics. The sector and seal concept involves use of standoff sensors to determine enemy locations and then deployment of isolation devices from afar to seal off the enemy from sensing or attacking our own forces. Nodal warfare also uses standoff sensors and weapons, but the main goal is to locate and disable key nodes such as C2 sites, communications relay platforms, power sites, water supplies, airports, and air defenses. These methods also extend current concepts such as ejecting entrenched defenders though use of comprehensive situation awareness and standoff weapons. One notion is to predetermine enemy egress routes and then flush them out in a desired direction. Thus, rather than playing into the strengths of an urban defense, these concepts "entomb" the would-be defenders, deny them infrastructure access, and force them into disadvantageous positions. The third concept, noncombatant control, uses technology to identify, tag, and move noncombatants out of the combat zones. In some cases, the inhabitants may even be enlisted to gather intelligence and coordinate movements. It is likely that implementation of these concepts may require expansion of the time required to complete a given mission, thereby resulting in a "reduced casualties for increased hours" tradeoff.

The underlying principle behind these concepts is that the main proportion of the urban fight may be carried out at standoff ranges, and that the enemy may be forced away from his favorable positions. Risky interior engagements should be relegated to unmanned air and ground systems as much as possible.

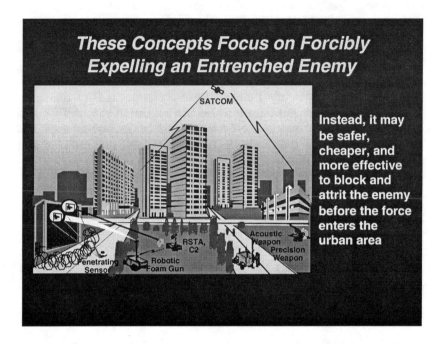

Application of these concepts to urban areas is illustrated here. Even in a relatively simple situation, such as eviction of a sniper team, there may be requirements for special sensors, nonlethal devices, command and control vehicles, and precision weapons. It may be easier to use many of these techniques on an approaching enemy rather than one already entrenched in a building.

This section describes how the three components can go together as an effective strategy.

**Expected Sequence of Events
For Prepped City Under Attack**

- With little warning, invading forces approach outskirts
- Overhead and ground sensors determine size, speed, and direction of force
- Obstacles are deployed, slowing and channeling force in desired directions
- Jammers, anti-materiel devices, smoke, local fire support used to further confuse and slow forces
- Enemy forces which succeed in penetrating city are detected and tracked; non-combatants cleared from area; local forces provide intel and reactive fires
- C2 centers coordinate precision fires from rapid reaction and standoff platforms, attriting stalled/slowed/ isolated enemy vehicles and infantry

An example scenario for a prepared urban area might start with a surprise attack from a nearby border. The enemy force is quickly detected by the prepositioned air and ground sensor net. Obstacles set in belts outside the city soon slow the attack to a crawl. Coordination of the enemy force is degraded quickly by jammers, smoke, and coalition rocket, cannon, and mortar fire, along with air attacks.

Nevertheless, some portion of the attack may succeed in the first few hours or days and part of the city might be occupied. The invaders are then tracked, slowed, and isolated using in-town obstacles. When the rapid reaction force from neighboring areas or CONUS is able to close, its fires are immediately directed from mobile and stationary C2 centers. Long-range naval, USAF, and ground fires are similarly targeted from the C2 centers. A portion of the ground force may be designated to clear any enemy strongholds in the city. This would be done using offset engagement.

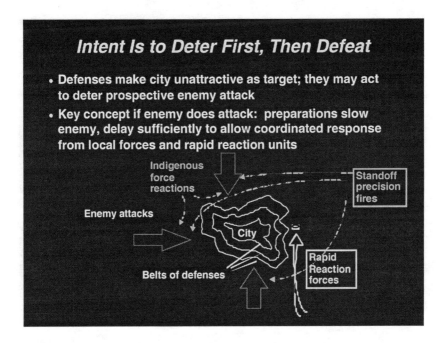

This graphic illustrates some of the process. Prepositioned defensive belts extend far from the city periphery, allowing precision fires from the city and from rapid reaction forces to concentrate on the slowed targets. These belts allow a coordinated fall-back with increasing canalization of the attackers.

Part of the key to successful defense outside city boundaries is the recent development of wide area (smart) mines and other new technologies. Smart mines can cover large areas, killing or slowing ground vehicles and low-flying helicopters. Higher-altitude air mobile operations may be stopped using increasingly capable man-portable and vehicle-mounted air defense systems. Software radios and tactical internets can coordinate indigenous, rapid-reaction, and long-range fires. Fast-setting foams can set up obstacles or can establish bridges over broken terrain. Robotic air and ground vehicles can resupply the defenders with additional sensors, obstacles, and weapons as they are needed. Ground forces equipped with fast-reacting weapons can engage fleeting targets, and standoff precision fires can go after key nodes and moving targets in the open.

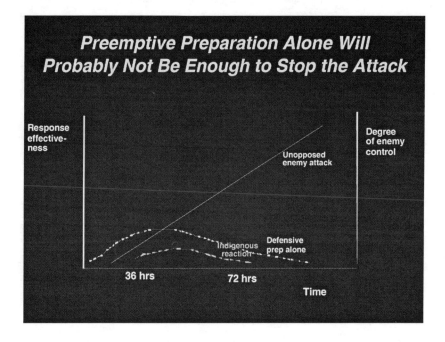

This is the first of a series of three slides intended to show the expected synergy between the three components. If unopposed, an enemy attack on a large urban area might take several days to occupy and defend the area. Indigenous reaction without any concerted preparation of the city might take some time to initiate a response, and then would likely have little effect on the outcome. Preemptive defensive preparations would have an immediate effect and would result in more enemy attrition and disruption, but without additional fires would probably also simply delay the occupation.

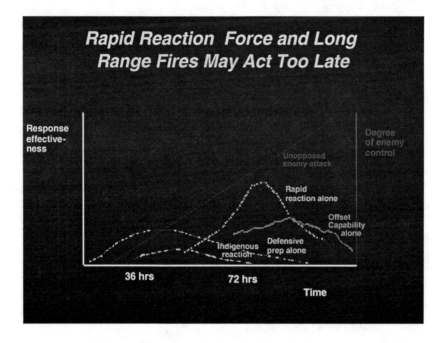

The effects of long-range fires and rapid reaction forces are also expected to be insufficient by themselves. These responses may arrive after much of the enemy force is already established in the urban area. The rapid reaction units would then be forced to root out the enemy forces at high cost. Long-range fires, even precision ones, could cause extensive collateral damage and noncombatant losses when targeted against enemy positions in the city.

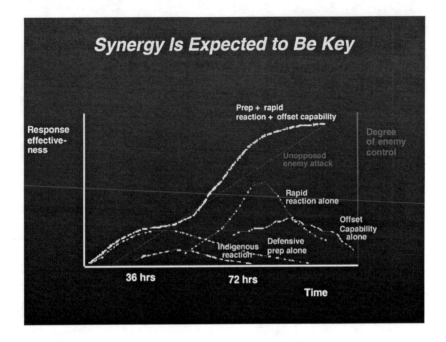

The effects of all three components are expected to be highly synergistic. Defensive preparations used in concert with rapid reaction
forces should work together to prevent entry, disrupt the enemy
attack, produce slow-moving targets for mid- and long-range fires,
and result in substantial attrition. Those enemy units that do achieve
a lodgment in the cities can then be isolated and evicted using offset
engagement techniques combined with limited numbers of close
combat operations.

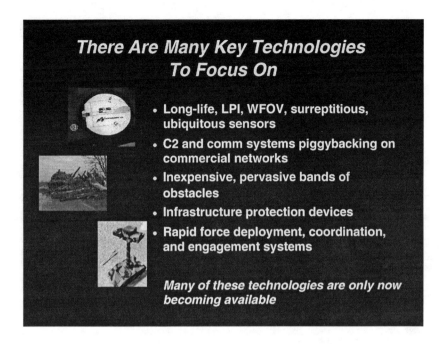

There Are Many Key Technologies To Focus On

- Long-life, LPI, WFOV, surreptitious, ubiquitous sensors
- C2 and comm systems piggybacking on commercial networks
- Inexpensive, pervasive bands of obstacles
- Infrastructure protection devices
- Rapid force deployment, coordination, and engagement systems

Many of these technologies are only now becoming available

Preemptive preparation of urban areas against attack is not a new idea. Seoul, Tel Aviv, Grozny, and many other cities have extensive preparations on avenues of approach. One difference between those traditional defenses and the concepts put forth here are the many new technologies that enable flexible, effective disabling of enemy attacks. Only in the last few years have new microsensors, mobile commercial communication systems, 3-D visualization tools, robotic security and resupply systems, special foams, and other technologies matured to the point where they can contribute to flexible, cost-effective defenses. Many of these systems can be shifted quickly to the specific areas being attacked, unlike historic defenses such as walls, ditches, minefields, and protective fires. They can also capitalize on the principles of massed fires, hitting an enemy's most vulnerable points, seizing the initiative with counterattacks, and using multimode, redundant systems that cannot be easily defeated by countermeasures.

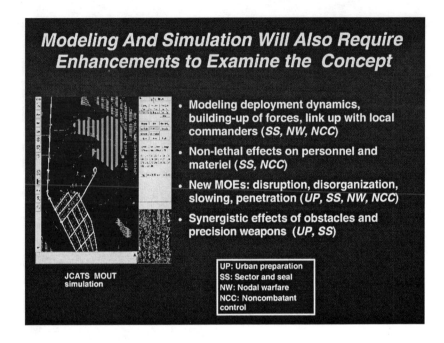

Modeling And Simulation Will Also Require Enhancements to Examine the Concept

- Modeling deployment dynamics, building-up of forces, link up with local commanders (*SS, NW, NCC*)
- Non-lethal effects on personnel and materiel (*SS, NCC*)
- New MOEs: disruption, disorganization, slowing, penetration (*UP, SS, NW, NCC*)
- Synergistic effects of obstacles and precision weapons (*UP, SS*)

JCATS MOUT simulation

UP: Urban preparation
SS: Sector and seal
NW: Nodal warfare
NCC: Noncombatant control

Replicating the complexity of these activities will require enhancements to traditional wargaming and simulation systems. Augmentations needed include representation of the interactions between indigenous, rapid reaction, and standoff forces, particularly with respect to how obstacles and mid- and long-range fires act synergistically. At the same time, post-processing must characterize many new measures of effectiveness such as incapacitation of personnel, disabling of vehicles, isolation of units, and impairment of command and control. Systems such as JCATS have some of these capabilities incorporated in their code, but much of this functionality will require new programming.

Outline

- Why preemptive MOUT?

- Basic components

- Concept synopsis

- Observations

The last section summarizes our observations and insights to this point.

Observations/Insights

- In many situations, U.S. will not be able to respond quickly enough to prevent occupation of urban areas
- Ejecting an entrenched enemy from a town is far more difficult and costly than preventing his entry
- Preemptive defensive measures can slow or disrupt invasion, but may not be enough in themselves to prevent loss of city
- C2 systems pre-wired into cities at risk must be able to control both indigenous defenses and external rapid reaction forces
- High resolution modeling and analysis needed to determine the most effective investment strategy—acquisition of obstacles, comm nets, C2 systems, weapons platforms, etc.—taking into account possible countermeasures

Many of the observations to this point are simple ones. It is becoming increasingly clear that the United States and its allies cannot respond quickly enough to stop an aggressive force from taking over many key urban areas. Going in and driving out an entrenched aggressor from a newly occupied urban area could be costly, even with dramatic new tactics and technologies. Preemptive preparation can buy time, but in itself is probably not enough to stop or deter an aggressor.

The approach with the most promise appears to be a combination of preemption, rapid response ground forces, and standoff fires. These disparate capabilities need to be orchestrated from a unified command and control center to achieve the greatest shock, delay, and disruption of the enemy attack.

The next questions concern the most effective investment and application strategy. How should scarce R&D and acquisition resources be parceled out? How much should the host country be required to shoulder? What agreements should be in place for best application of the system? Should control of the response be with the host country, a coalition, or the joint task force? Some of these questions may be answered though high-resolution simulation and some through analysis and negotiation.

JOINT MOUT MISSION AREA ANALYSIS
AND MISSION NEED ASSESSMENT
LtCol Duane Schattle, USMC

JOINT MOUT
Mission Area Analysis and
Mission Need Assessment

LtCol Duane Schattle
J8 LLWAD
695-4657

It's always interesting to come here. One of the first things you have to do is establish your credibility. I'm always amazed. I've heard General Scales a number of times and the guy's got impeccable credentials. He is a general officer in the Army. He's a Ph.D. He just commands respect immediately. Then you've got other guys like Dr. Andrews and of course Randy Steeb and his fourteen years of education. I had fourteen years of graduate education, too, and my parents were really glad when I finally got that high school diploma. One of the things I have to do is establish my credibility right up front because I'm wearing a Marine Corps uniform and I'm wearing a Joint Staff badge. A lot of people say, "you work in urban issues and that's really the Commandant of the Marine Corps thing and you're really probably a spokesman for the Commandant of the Marine Corps." I have to tell you, it's really not true. I really am a joint player. I really do believe in the joint world, and to establish that as fact I'll tell you that my past goes way beyond my experiences in the Marine Corps. My dad was enlisted; he happened to have served in the Army, Navy, and the Air Force. He retired from the Air Force as an E-9 Chief.

What has the Joint Staff done in the past? Where are we going in the future? How do we view this whole issue of urban operations?

JOINT MOUT STUDY
A BIT OF HISTORY

- **LLW JWCA Review**
  - Popular Literature (Kaplan, Keegan, Kennedy, Tofflers...)
  - United Nations - Habitat II & Cairo Conference
  - World Resources Institute - Urbanization Statistics & Models
  - RAND - Urbanization Warfare Studies (Taw, Hoffman, Davis)
  - Service School Studies
  - Chechnya-Professional Literature - Intelligence Data
  - Defense Science Board (86, 94 and 96)
    - MOUT ACTD (1997-2002) & Urban Warrior (1997-1999)
  - Grozny (95), Chairman's (JCS) Concern
  - Joint Strategic Review (96)
  - MOUT Seminar Wargame (CSEEA) (96-97)
  - QDR (96-97)
  - Joint Staff Urban Capabilities Assessment (97)
  - Joint Warfighting Science and Technology Plan (98)

I want to tell you right up front that we weren't convinced that we needed to look at urban operations. So what we did was a front-end assessment to figure out if operations in cities are important. As General Scales mentioned yesterday, he's very critical of people that say just because the world is urbanizing we're going to go fight in the cities. I think that's probably an appropriate conclusion; just because the world is increasingly urban doesn't mean we're going to fight in the cities. There's got to be something else. So we looked at as much information as we could over a period of a year to establish whether or not this was a credible area for us to be concerned with. You can see some of the things that we looked at. Each one of these things involved lessons to be learned.

JOINT MOUT STUDY
Front End Assessment

- **Battles Examined (97)**
  - Stalingrad (42-43), Manila (45), Seoul (50), Hue (68), Beirut (82), San Salvador (89), Panama City (89), Mogadishu (92-93) and Grozny (94-95)
- **Demographic Trends**
  - UN: 45% urbanization today, 50% by 2005, 61% by 2025
  - DIA: Urban Warfare asymmetric threats likely to US forces
  - JSR (96): there will be more urban terrain in 2010
  - CPR (98): Doctrine of urban avoidance will not be possible
- **Political Landscape ⟶ Strategy of Engagement**
- **Joint Vision 2010**
- **New Warfare Reality**
  - Collateral Damage
  - Casualty Aversion
- **CNN**

We examined nine separate battles picked because we wanted to determine some general trends. Why these battles? Well it's because the people who did the study picked these, we approved them, and we wanted to see if there were any trends. Now the thing that came out of that study that was most alarming was the differences between what happened in Stalingrad in 1942 and 1943 and what happened in Grozny in 1994 and 1995. That really got my attention. When the Germans went into Stalingrad, millions of people died. And the Germans lost. Fifty years later, when the Russians went into Grozny, Chechnya, many attackers died and the Russians lost. Just like Stalingrad. The parallels between the things that happened in Stalingrad fifty years earlier and what happened in Grozny, even with all the technological innovations that the Russian Army had in 1995, were striking.

We looked at the demographic trends, but we also looked at more than the demographics. We looked at the DIA assessment. We looked at the JSR, the national defense panel. We looked at the Defense Science Board and what they had to say. We looked at the political landscape, including the fall of the Soviet Union and how things had changed. We looked at *Joint Vision 2010*. How are we going to do things differently and more efficiently in the 21st century? Then we looked at new warfighting realities. There are new realities that we have to be aware of, one of which is the collateral damage issue and casualty aversion. And casualty aversion is more than just concerns about friendly forces. It is also the noncombatant issue, and I think there is an argument that can be made after the highway of death incident during the desert war that we're even averse to killing too many of the bad guys. It affects our policy. So those are the types of things that we included in our front-end assessment. And then, of course, there's the CNN factor. We all know about the CNN factor. You're going to see it unless the enemy kicks out the press, as was the case in Kosovo. You're going to see live feeds as occurred during the Los Angeles riots with Reginald Denny, the guy who was dragged out of his truck. What kind of impact is that going to have? I suggest to you it's going to be like instant replay in the NFL. The American populace is going to immediately judge the effects and the value of our policies based on what they see on television. We have to pay attention to that. So we've got to do things smarter and better.

JOINT MOUT STUDY

- **Findings:**
  - **Not All or Nothing**
  - **Must maintain other capabilities**
  - **Our Success has created the problem**
  - **Need to address Joint Urban Operations Capabilities**

- **Concerns:**
  - **Avoid and lay siege**
  - **Destroy and plunder**
  - **Sensors-R-US**

We have never said that this is an all-or-nothing approach. If you listen to some people, they would suggest to you we don't need to invest money in urban operations; we don't need look at these types of things because they take away from other needs. We're suggesting that we take a balanced look at everything. We are in the cities already today; marines, soldiers, and airmen are operating in cities as we speak. It can be argued that our capabilities in open terrain have driven our enemies to the cities because they know they can't compete with us on the outside. And that's why a lot of these arguments about getting him before he gets to the city are great. If you can get him before he gets into the city, then good for you. But I think there are a couple of examples from the last ten years when we couldn't get him before he got into the built-up area. In Desert Storm, we had intelligence that the Iraqis were massing at the border headed for Kuwait City. Now General Scales would suggest to you that in the Army After Next war game, the National Command Authorities waited too long to commit American forces to kill the enemy. My question to General Scales is, What's new about that? We're a representative democracy, and democracies don't go to war quickly in most cases. It takes time. It takes time to build coalitions. We tend to hesitate before risking involvement in long, drawn-out operations. It's part of our way of doing business. But to suggest that the National Command Authorities are going to attack just because the generals say we should attack isn't consistent with history. Kuwait is a perfect example.

I suggest to you that this latest round in Kosovo is another good example. The enemy had forces in the immediate vicinity. The cities were very near by. How long does it take him to get from here to here, to move the people out, to take the cities, and how much force did we have to immediately influence what he can do? We didn't have enough force. So again, General Scales' argument, although it's a good academic argument, flies in the face of reality.

So, in our assessment we've said that we need to address the issues of urban conflict because we're probably going to be involved there in the future just as we are today. The concerns that we have include possible approaches. These were answers given to me when we were doing this front-end assessment. First of all, we're going to avoid fighting there. We're going to go around or we're going to lay siege to the city. But then when you ask the question: "Are you really going to lay siege to a city of a million people?" You're going to starve out the bad guys? Well, are the bad guys going to let the civilians eat or are the bad guys going to eat? Who's competing for food? Is the idea to set up camps where you can feed them? Are they going to let them go to those camps? And how are you going to feed a million people? Are we going to set up all the logistics first and then say, "Okay, here's where the camps are," and then send the military to protect the camps? By that time the enemy will be attacking those camps. I don't follow that logic. That will not work. We're not Alexander the Great, not Attila the Hun; we don't fight war that way. We believe in just wars.

Destroy and plunder is another approach. We're going to line up our artillery; we're going to line up our armor; we'll put our tanks and everything else out there. We're going to bomb the city into submission and then when we're all done bombing we're going to send the infantry in and let them clean it up. I think they tried that in Grozny. It didn't work. They tried it in Stalingrad. It didn't work. You are talking about linear combat, committing masses of divisions that the city will eat overnight. Is that what we want to do? And don't forget CNN. They're watching all this.

And then there's the Toys-R-Us option. We think we're going to have all of these technologies and we're going to throw them out there and we're going to know everything. The next question is, "Then what?" Well, we know everything. We have perfect situational awareness. Do you still go in? If you don't have the right capabilities to do something, you're still stuck. We don't want to end up in the 21st century doing what mankind has done throughout history.

JOINT MOUT STUDY DESCRIPTION

- <u>PURPOSE:</u> Conduct a JROC & Defense-directed Joint Mission Area Analysis and Joint Mission Need Assessment of U.S. Joint <u>Operational Warfighting Capabilities</u> in the Urban Environment.

- Defense Guidance
    - Continue studies to assess & explore Joint capabilities
    - Assess alternative approaches to conducting MOUT
    - Recommend actions to address doctrine
    - ID M&S & training requirements
    - ID requirements to achieve dominant MOUT
        - Include NLW
        - Build a roadmap to 2010 to find alternative approaches

So what we did was convince the JROC that we wanted to take a look at this. The secretary quickly established this as a Defense study. The guidance was to accomplish the tasks shown above.

JOINT MOUT STUDY
JUWG Formed (1998)

- OSD
- State
- CINCs
- Services
- Joint Staff
- DOD Agencies
  - DIA, NIMA, NGIC, JWAC, NRO and ALSA
- 90+ personnel (internet)
- Weekly meetings
- Monthly Offsite

We formed a joint urban working group with members from OSD. We're still working on the State Department; Ambassador Oakley has been involved because this issue is much broader than military only. It is an interagency process. That's one of the problems we're having; we can't get the State Department formally involved. We're going to continue to try to do that, but I've got to tell you it's very difficult. You can see the other organizations that are involved. All the services have a speaking part, as do the CINCs. We meet weekly for about four hours to either bring in people to talk to us, take a look at technologies, or to discuss issues, then monthly we do the same types of things for an entire day.

JOINT MOUT STUDY DESCRIPTION

- **PHASE I (3 PARTS):**
  - Assess the need for Joint Operational MOUT Doctrine;
  - Conduct a survey of Analysis and Training M&S Tools for Joint MOUT operations;
  - Identify Joint Operational Mission Needs for MOUT.

The urban environment is not like the elephant with people trying to take one bite at a time. It's more like a herd of elephants and they're all running. The objective a year and a half ago was to slow them down and then get all the services and all the intellectuals involved in taking bites of these elephants so we can digest the subject over time. We identified these three areas that we thought were the most important. We wanted to take a look at the MOUT doctrine issue. We wanted to look at models and simulations, and we also wanted to look at mission needs for the joint force commander. The focus for us was at the operational level and joint force commander level rather than the tactical level.

**PHASE I: JOINT MOUT DOCTRINE
ASSESSMENT... COMPLETED**

- **Findings:**
  - **Current Doctrine Inadequate**
    - **Tactical, Ground-oriented, Linear and Attrition Driven**
    - **Not Joint, not Combined, nor Interagency**
    - **Operational Level Void**

  - **Needs**
    - **Joint Operational Doctrine**
    - **Joint Operational Concept**
    - **Short/Near-term Guidance**

Back in 1995 the Chairman of the Joint Chief of Staff, General Shalikashvili, turned to the Joint Staff after watching what had happened in Grozny and asked, "Do we need to pay attention to this? Do we need joint doctrine for MOUT?" The Joint Staff came back and said no, the Marines and the Army basically follow the same doctrine, and if we made a joint doctrine we'd just be taking what they've already said and putting it under another cover. Well, the problem is that Army and Marine doctrine isn't joint. It doesn't bring all the aerospace power and naval pieces together. We did an assessment. Ambassador Oakley was part of the team. We took former JTF commanders, including General Johnson from the Marine Corps, who was in the JTF in Beirut and in Somalia. Of course Ambassador Oakley was in Vietnam, Beirut, and Somalia. General Horner from the Air Force, General Otis, and General Downing, all retired generals, were asked these questions. Then we took it to the urban working group. We found that urban doctrine was tactical; it was ground-oriented; it was linear; and it was attrition-based. It would lead us down the same path as the Russians followed in Grozny if we didn't do something. It wasn't joint. It wasn't combined. It wasn't interagency. And there was a huge operational-level void. Putting it simply, the group agreed that we needed joint operational doctrine. We also needed an operational concept around which to build the doctrine, and we needed short-term guidance because writing the doctrine could take as long as two to three, maybe even four years.

PHASE I: JOINT MOUT DOCTRINE ASSESSMENT... COMPLETED

- **Actions Taken (Urban Working Group Initiated):**
  - JMOUT Doctrine (2001)... USMC Lead, USA Review Authority, J-8 Sponsorship
  - Ops Concept ...(w/Doctrine)...USMC Lead, UWG Guided
  - JTF Handbook (1999)... USAF Lead, UWG Guided

The urban working group is the joint doctrine sponsor, with the Marines leading the writing effort and the Army acting as the reviewing authority. The operational concept was developed through a series of war games. The JTF handbook is being written by the Air Force to fill the void until the formal doctrine is completed.

PHASE I: JOINT MOUT M&S ASSESSMENT...COMPLETED

- **Findings:**
  - JWARS and JSIMS will not address MOUT in near-term and long term is questionable
  - Training Tools... Appears Adequate But it Lacks Urban Terrain Data Bases
  - Analytic Tools Are Inadequate
  - Needs
    - Analytic Tools (Tactical and Operational) For:
      - Capability Assessments
      - Campaign Analysis
      - Rehearsal Tools
    - Terrain Data Bases for Both Analysis and Training

Models and simulations. We found that it is very difficult to do anything other than qualitative analysis because, as has been mentioned by Randy Steeb, there's really nothing out there for us to use. We found that out in the QDR. We can measure how B-2s and B-1s perform. There are models to do those types of things. When we got to the urban setting, what did we have to do? We had to sit around in a group and talk about it. That's good, but it doesn't carry the day when you talk about spending money in DoD. It doesn't give you credibility. We found that nothing has changed in those 30 years. That ought to be alarming. It surely was to us. We found that a lot of people doing different types of things to make training realistic, but there is a serious database issue. We've got to address that.

We need analytic tools. The few that are out there are not very good. We need to be able to do campaign analysis. We need to have rehearsal tools, just like for any other environment. And we need terrain databases for both analysis and training.

PHASE I: JOINT MOUT M&S ASSESSMENT...COMPLETED

- **Actions Taken:**
  - **JCATS identified as possible short-term solution and long-term bridge; requires:**
    - V&V (JWFC, $500K)
    - Identifying Acceptability Criteria... What Should Model Do (UWG Can Help)
    - Terrain Data Bases (larger than normal data requirement)
    - FY 2001-2005 Defense Language (Apr 99) states:

    "Critical to this effort, the Department, through the Joint Staff, must develop short-term analytic tools to assess those capabilities."

Then we stumbled on JCATS. Although it's not the panacea, it is, in our judgment, the answer in the short term. We found that JCATS has a lot of the things that allow it to be that link between present and future capabilities. We'd have to identify the JCATS acceptability criteria operational level, and then each of the services could do whatever they think is necessary. All the services are using JCATS right now, but the back of a terrain database keeps coming back to haunt us. We need quantitative analysis.

PHASE I: JOINT MOUT MISSION NEEDS COMPLETE

- **Findings:**
  - C4 and ISR require independent detailed assessment now
  - Lack of M&S tools makes assessments difficult and leads to qualitative not quantitative results
  - Because of closed terrain and population, etc., often tactical concerns become operational issues
  - No centralized focus to address Joint Requirements
  - Much more needs to be done... (Phase II & III address)
  - Current study identified 108 Candidate Mission Needs

The third piece was mission needs at the joint force level. We immediately identified that we were not going to be able to address certain things. We knew that we needed a C4 and an ISR study and thus initially chose to focus on just those two areas. These were not decisions that are made just because somebody in our shop thought it was important. The Urban Working Group came up with these recommendations. The J6 is doing the command and control part of the study. It's ongoing right now with help from the Decision Support Center, which works directly for the Vice Chairman. And the J2P is doing the ISR piece.

Another thing we found out was that the lack of models and simulations make these studies difficult. We're trying to do a work-around with OPNET, JCATS, and other models, but it's very difficult to do. It just reinforces the fact that we need help in the simulations area.

Another one of our findings as we looked at joint mission needs was that closed terrain often causes tactical issues to become operational concerns. That may not sound like a very big deal to you, but it is a huge deal when we talk about work on the Joint Staff that starts getting into the service arenas. The lines between tactical, operational, and strategic are beginning to blur. Let me give you a couple of examples. Normally you would think body armor is a tactical concern. That's an issue for the individual on the ground. And it is an issue for the individual on the ground. But if you don't look at it from the joint perspective, this is what you end up with. The joint force commander has to operate in an urban environment with two types of ground force—let's say the Marines and the Army. He's got to employ ground forces in a high-risk situation. The National Command Authorities then say, "We're really concerned about force protection issues. I want to make sure those guys have the most protection as possible." Then somebody comes to him and says, "Well, general, the Army guys have level 3 body armor, the Marines only have flak jackets." You think that's a tactical concern for the JTF commander? I submit to you that it's not. Here's another example: combat identification. Combat ID is normally a tactical concern; we're not too concerned about that at the JTF commander's level, right? I guarantee you it's a concern for the JTF commander, especially when you're in a city where you can't see 15 feet around a corner and friendly force casualties could cause mission failure. That's what we're talking about. A lot of these things have operational impact, and it's important to note that.

We also found that there's no centralized focus for addressing these joint requirements; people are working things on their own. We also found that we're only scraping the tip of the iceberg right now and there's a lot more that needs to be done. What we've done is lay out a plan to try and address those needs.

PHASE I: JOINT MOUT MISSION NEEDS COMPLETE

- **Actions Taken:**
  - **UWG Narrowed 108 Needs to Top 18**

Communications	Rapid & responsive firepower
Precision effects	Population control
Knowledge of cities	Mobility
Threat detection/ neutralization	Consequence management
JTF single common picture	Medical
Information control	Logistics
Precision navigation	Training
CID	Training analysis
CSAR in urban	Campaign analysis

There are the top 18 identified needs that are to be addressed in a phase 1 study. Some of them look very familiar to you. What we're going to do is take a look at existing mission needs statements and see which of these areas are covered for the urban environment. For those that are not, we'll make recommendations to the JROC and the DPAG regarding possible solutions. We ought to finish these 18 off in the next three or four months, then go back to the JROC and tell them we'll start going after those remaining out of the 108.

PHASE I: JOINT MOUT MISSION NEEDS (ACTIONS TAKEN, CONT'D)

- **Phase II Tasks Identified:**
  - **Follow-on Mission Needs...**
    - Refine and prioritize Candidate Mission Needs/forward to JROC
    - Input C4 (J-6/DSC) and ISR (J2/DSC) Study Mission Needs
  - **Transition Study...Develop**
    - Concept Exploration/Roadmap
    - Institutionalization of Joint Mission Needs Options
    - MOUT Oversight Options
  - **Address Terrain Data Base Issues**
    - Prioritization
    - Sponsorship
    - Funding, etc.
  - **Address Future Joint and Service Studies**

These are the things we are doing in phase 2 for this fiscal year. We will refine those 18 and take a look at what comes out of the C4 and the ISR studies. We'll add them to this as we take it forward and rebrief it. We're also doing what's called a transition piece. We've been directed by the secretary to build a science and technologies roadmap out to the year 2010 to bring all of these concepts together. We will then have a vision of what the future's going to be and what our experimentation plan will be. We're going to look at how we institutionalize joint mission needs and joint requirements. And then we're going to provide options to the secretary and JROC on how to best manage this. Should it go to ACOM? When should they get it? In the meantime, what do we do? Do we give it to a service? Should it be a joint program office? We want to answer those questions and provide options so the decisions can be made.

We're also going to address the terrain database issue. Now, we're not going to do an in-depth study in the terrain database, but the urban working group will address things like prioritization. If we're going to build terrain databases we have to establish priorities. Is it going to be large cities, small cities? Is it going to be five different cities so that you can do some type of assessments? These are not trivial questions. We are talking about C4ISR capabilities in cities. Take New York City for example. We can build a terrain database of New York City, but New York City is flat. It's got a lot of high-rises, but it's flat. You go to San Francisco, a lot of high-rises, and it's on hills. Most places in the world don't have all those high-rises. So we have to make some choices here because there's only so much money available.

We're also going to address the future of joint and service studies, to establish what is it that each of the services feels that they should be doing in this environment. We'll then attempt to get them to sign up for that.

PHASE I: JOINT MOUT MISSION NEEDS
(ACTIONS TAKEN CONT'D)

- **Phase III Assessment Priorities Identified:**
  - JMOUT Training and Facilities
  - Urban Information Architecture
  - JMOUT Center of Excellence/Analysis
  - Future Joint and Service Studies (Precision Effects, Combined Arms, Strike, Airpower, Artillery, Information Systems, RW, Armor, Space, Logistics, etc.)
  - Data Base Study
  - Others as Identified...

That was phase 2. That's where we are now. Phase 3 is for next year. At the very top of the list you see joint MOUT training and facilities. It's a huge issue. We just briefed the JRB and recommended that they support accelerating this study. It was approved by the JRB. I assume it will be approved by the JROC and the CINCs and we'll start down this road in the next three or four months. What are we looking at? We don't have places to train. I know that the Marines did the best they could, had the mayor agree that they would be able to land across the beach in Monterey, but they didn't get to make the amphibious landing. The same thing happened in San Francisco. This is a huge problem. What we don't want is to have the services stepping out on their own and building millions of dollars worth of small MOUT facilities with no strategic plan. We'll approach this issue the same way we approached the others. What are the requirements for urban training and training facilities? What do we have now, and what's the gap? We will provide options. Do we need a national training center for urban operations; a huge city built in the desert? How much is that going to cost? And how many people are going to use it? Do we need to expand the JRTC? Is that a better option? What are the best options to meet the requirements?

You can see the other areas we're going to look at next year.

JOINT MOUT MISSION NEEDS
PHASE I .....COMPLETE

- **Reoccurring Themes**
  - **Joint capabilities required (ISR, C4, CAS, etc.)**
  - **The city is more than an environment, it is a living entity**
  - **The lack of analytic tools makes quantitative MOUT analysis almost impossible**
  - **In urban areas the distinction between tactical and operational concerns is often unclear**
  - **There is a need for a single focal point to address MOUT**
  - **The term MOUT inhibits progress**
  - **Much more needs to be done to address Joint MOUT capabilities and needs**

These are the recurring themes that keep coming back to haunt us. Joint capabilities is a joint fight. It's not a single service fight. It's not a land fight. It's not an air fight. It's a joint fight. It's interagencies. The city is more than an environment. It includes the most difficult environment because of the three-dimensional issues. But the real factors that make it different are the civilians in that city, and the infrastructure that ties it together. It is more than just an environment. The urban working group likes to compare the city to a living organism. In order to operate successfully, and remember we're not just talking about fighting, a force must remove the cancer without killing the patient. But there is a time issue. You can't let the cancer grow to the point where it is so advanced that the patient's going to die anyway. In addition, you've got to make sure you're cutting the cancer, that you don't make a mistake and cut off the wrong part. If you're doing brain surgery, make sure you do the right part of the brain so you don't unnecessarily affect other parts of the body.

The term "MOUT" itself inhibits progress. Why? What is MOUT? I bet that 80 percent of you would say, "That's marines and soldiers on the ground fighting." But that's not the definition of MOUT. Mark Sumner pointed out yesterday that the Army and Marine Corps have basically the same definition. It's much broader than combat only. It's all operations in built-up areas. So the urban working group has decided it is not going to refer to the term "MOUT" in the future. We're going to use the term "urban operations."

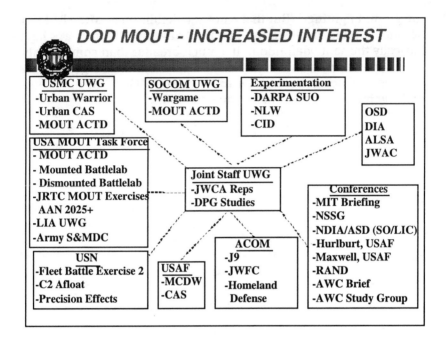

The reason I put on this slide is to show you all the things that are going on in the urban area now; there are probably some things missing. Two years ago there were really only two initiatives on-going: the MOUT ACTD and Urban Warrior. There were some RMA net assessment games that were tied to those two. Since then, you can see all these other things that have begun: the use of minimal collateral damage weapons, various studies, Army After Next urban initiative conferences; all of these types of things are now ongoing. And the urban working group is a key part of that growth.

URBAN WARRIOR AND USMC URBAN OPERATIONS
Col Gary Anderson, USMC

Urban Warrior

Col Gary W. Anderson
Chief of Staff
Marine Corps Warfighting
Laboratory

This book outlines some key problems of urban combat:

- Urban Navigation.
- Urban Communication.
- Civilians intermixed with fighters.
- Cultural Intelligence.

What is the Marine Corps doing about it? One example is the Squad Radio solution.

A Critical Question

"How is it that a nation that could land an unmanned little go-cart on the surface of Mars couldn't steer a convoy five blocks through the streets of Mogadishu?"

Mark Bowden

Blackhawk Down

D:/col anderson/brief 7apr99

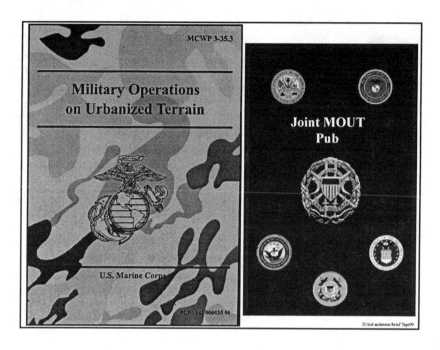

MCWL used the current doctrine on the left as the baseline. We hope to contribute our lessons learned to emerging joint doctrine on the right.

The New Warrior Class

The enemy of the future (Hezbollah youth) will be:

- 25 in the year 2020
- Innovative and skilled in urban combat
- None have been captured alive in seven years
- Kill rate for the Israelis is 10:1 in his favor
- He has cousins ranging from Africa's Gold Coast to East Timor

It is important to understand the environment and the culture.

DRAGON DRONE

- Dragon Drone is a Close Range Tactical UAV that has become the tactical UAV testbed for the Marine Corps. It is currently deployed with the 15th MEU in order to test its ability to extend the area of influence of a modest forward afloat expeditionary force.

- Primary mission is reconnaissance, surveillance, and target acquisition.

- Is also utilized as a payload development vehicle.

FY99 Operations

- Complete 15th MEU deployment in Dec 98.

- LSD-44 shipboard installation by Dec 98 for 26 MEU.

- 26 MEU deployment - Apr 99.

- Continued payload development for future incorporation into Dragon Warrior.

Dragon Drone is a first-generation short-range UAV. It allows us to see down the block and around the corner. It is a surrogate for more advanced systems, but it has been so popular that we are actually deploying it aboard our MEUs afloat until Dragon Warrior comes on line.

Surveillance and Reconnaissance Ground Equipment (SARGE)

Surveillance / Reconnaissance

•BSTF (Battle Shaping Task Force)
– Norfolk, Summer 1998

•LOE3 (UW)
– Camp Lejeune, Summer 1998
– Free Play, MOUT Facility

•LOE3 (CW)
– Virginia, Nov 1998
– RSTA (1)
– RSTA (2) 4th LAR

Another example of technology that can help.

Combat Decision Range

- Train NCOs in combat decision making and leadership

- For use by every infantry regiment in the Marine Corps

- Operational March 1999

- Three-man mobile demonstration team

- Computer-based training system

- Uses seven major conflict/combat scenarios

- "Trains the trainers"

- Three days of instruction

• Extremely successful training capability

• No hard data on performance improvement, but universally hailed by commanders and trainees as a desirable product

A program to train squad leaders.

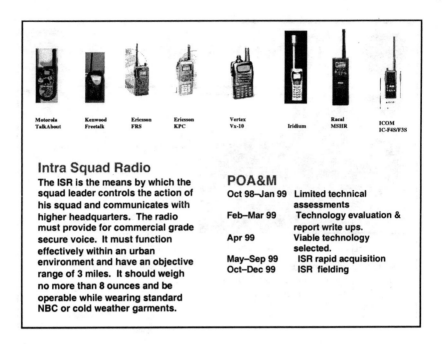

Motorola TalkAbout   Kenwood Freetalk   Ericsson FRS   Ericsson KPC   Vertex Vx-10   Iridium   Racal MSHR   ICOM IC-F4S/F3S

Intra Squad Radio

The ISR is the means by which the squad leader controls the action of his squad and communicates with higher headquarters. The radio must provide for commercial grade secure voice. It must function effectively within an urban environment and have an objective range of 3 miles. It should weigh no more than 8 ounces and be operable while wearing standard NBC or cold weather garments.

POA&M

Oct 98–Jan 99	Limited technical assessments
Feb–Mar 99	Technology evaluation & report write ups.
Apr 99	Viable technology selected.
May–Sep 99	ISR rapid acquisition
Oct–Dec 99	ISR fielding

- Technology at work.

- We are working to provide better communications and situational awareness to the individual soldiers and marines.

- We are looking into the possibility of getting commercial off-the-shelf radios and GPS down to the squad level, perhaps even to the individual marine, in the near term.

- We are also looking at getting a common, tactical, over-the-horizon picture that will include icons that actually interact. This has the potential to decrease the number of personnel who are currently often engaged in mundane staff activities such as posting unit locations.

- Other solution sets—nonlethals.

- Perhaps the biggest winner.

Boom Gun

- Crane that has remote control .50 caliber machine gun (with camera) at end of boom
- Uses same weapon mount as Mobile Counter Fire System
- Used for situation awareness and force protection

- Proved to be a very good idea
- Technology used was not sufficiently developed to ensure adequate reliability
- Requires more work on tactical employment to fully assess

Technology capability that can be useful in an urban environment.

Force Protection Clothing

- Experimental Urban Camouflage Uniforms
- Reversible Urban Uniforms
- Thermal Thellie Suits
- Improved ability for marines to avoid detection on the urban battlefield
- Improved load bearing capability
- Improved ballistic protection for individuals

• Urban camouflage pattern proved to be very successful

• Improvements are indicated for protective accessories in order to enhance wearability

Portable Reverse Osmosis Water Purification System (PROWPU)

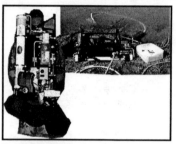

– Man portable system

– Capable of supporting 42-man infantry platoon (250 gallons)

• Capability judged to be a promising idea for the support of isolated small units

• Portability requires improvement if intended as a man-packed item—too heavy!

This is deployed with MEUs today.

MV-22 Compatible Tactical Vehicle

- Assault variant provides a promising solution to the helicopter-transportable vehicle requirement
- Met all mission requirements during AWE
- Logistics variant not used due to its size (not internally transportable)
- Requires additional experimentation to provide valid assessment of tactical utility

– Internal and external MV-22 compatible

– Highly agile, extremely stable on all terrain

– Can carry 3000-lb payload (four personnel and equipment)

– 2 variants:
  · Light Strike Vehicle
  · Light Tactical Vehicle

Another winner. This may not be the vehicle chosen, but we like the capabilities.

Multi-lingual Interview System

– Hands-free operation using voice-recognition

– Resulting native-language speech can direct action or obtain a "yes or no" response

• Technical success during AWE both as a means to deal tactically with local populace and for enhancing medical care to foreign language speakers

• User interface aspects like the audio output and controls require refinement

Aviation LTA

9 – 16 June 99

Yodaville

- Laser Guided Training Round
- Maverick K EO PGM
- Inert TOW
- Rapid Targeting System
- Tactical Airborne Target Acquisition System (H-1 surrogate)
- CUTLASS Lethal UAV (H-1 surrogate)
- Dragon Fire 120mm Mortar
- Dragon Drone UAV
- Pioneer UAV
- K-Max Helicopter
- LADAR (Laser Radar)

In March 1999, the MCWL conducted Urban Warrior. The primary focus was on C4ISR and ground/logistics elements.

In June 1999, the MCWL will conduct an Aviation Limited Technical Assessment at Marine Aviation Weapons Training Squadron-1 in Yuma, Arizona.

Aviators training in the urban environment is critical. Yodaville provides an urban training environment for aviators conducting close air support missions. The next step will be to integrate this range with ground combat forces.

MCWL: Quo Vadis?
X-Files

X-File 3-35.1	X-File 3-35.2	X-File 3-35.3	X-File 3-35.5	X-File 3-35.6	X-File 3-35.7
Urban Attacks	Combat Squad Leader	Battle Captain	Urban Defense	Urban Patrolling	Security Operations

Military Operations on Urbanized Terrain (MOUT)	Military Operations on Urbanized Terrain (MOUT)	Military Operations on Urbanized Terrain (MOUT)	Military Operations on Urbanized Terrain (MOUT)	Military Operations on Urbanized Terrain (MOUT)	Military Operations on Urbanized Terrain (MOUT)
USMC Warfighting Lab U.S. Marine Corps	USMC Warfighting Lab U.S. Marine Corps	USMC Warfighting Lab U.S. Marine Corps	USMC Warfighting Lab U.S. Marine Corps	USMC Warfighting Lab U.S. Marine Corps	USMC Warfighting Lab U.S. Marine Corps

X-File 3-35.8	X-File 3-35.9	X-File 3-35.10	X-File 3-35.11	X-File 3-35.12	X-File 3-35.13
Combined Arms	Battalion and Below Communications	Directed Energy Weapons	Humanitarian and Disaster Relief	Urban Sustainment	Tactical Instrumentation
Military Operations on Urbanized Terrain (MOUT)	Military Operations on Urbanized Terrain (MOUT)	Military Operations on Urbanized Terrain (MOUT)	Military Operations on Urbanized Terrain (MOUT)	Military Operations on Urbanized Terrain (MOUT)	Military Operations on Urbanized Terrain (MOUT)
USMC Warfighting Lab U.S. Marine Corps	USMC Warfighting Lab U.S. Marine Corps	USMC Warfighting Lab U.S. Marine Corps	USMC Warfighting Lab U.S. Marine Corps	USMC Warfighting Lab U.S. Marine Corps	USMC Warfighting Lab U.S. Marine Corps

MCWL synthesizes information derived from experiments and quickly gets it out to the operating forces for use and feedback. These publications, while not doctrine, are used as training enhancements for the operating force. Much of this material will go into or become Marine Corps doctrine as appropriate.

MEDICAL SUPPORT FOR URBAN OPERATIONS
COL Lester Martinez-Lopez, USA

Medical Support for Urban Operations

COL Lester Martinez-Lopez

I would like to discuss with you what it takes to conduct the medical mission in urban operations. In my experience, all too often the medical support requirements are an afterthought. So, I would like to focus today on the planning and execution of the medical mission in MOUT.

Pre-Deployment

- Intelligence Preparation of the Battlefield (IPB)
  - Get and study the plan or the order
  - AFMIC Products
    - Country Study/Classified Studies
  - Internet
    - USAID Reports/Country News/NGO's (ICRC)/WHO
  - Country Military Group/Embassy

Upon receipt of a deployment order, it is critical to study carefully the OPLAN or OPORD. One needs to note, in particular, the nature of the mission as it relates to the medical support required by the force and by the civilian population. For example, in Hurricane Mitch the main effort dealt with medical and engineering support to the disaster victims, whereas in Haiti, the mission was to establish conditions to allow for free elections to be held and to stabilize the country.

Questions to address include: Are there any special restrictions in terms of the medical care to be provided? Is the mission to give care only to the deployed force, or is medical support to the civilian population also encouraged? Is there a "nation building" intent or a large refugee population to be cared for? Each will require a different set of packages and medical planning. Other inputs include force and civilian casualty estimates and whether one will be supporting a joint or combined operation.

Next, one should obtain and study all of the country's and that particular city's medical and intelligence data. The sources may be classified for security reasons, as are these from the Armed Forces Medical Intelligence Center (AFMIC). However, probably the most up-to-date information will come from the Internet and such organizations as USAID, the country's newspapers, NGOs, WHO, and the ICRC, among other sources. These will provide additional information on medical issues such as disease outbreaks, the status of the health care infrastructure, and the like.

Also, always consider talking with the country group beforehand and with personnel both at the embassy and with other military groups, such as foreign military officers studying at one of the U.S. war colleges. If available, they can help inform you about events in their own country and identify points of contact for you beforehand.

Pre-Deployment

- **IPB**
  - Country Summary
  - Weather and Environmental Threats
  - Disease Threat and Countermeasures
  - Medical and Sanitation Infrastructure
  - Terrorist or Enemy Threat

To continue with the discussion of Intelligence Preparation of the Battlefield (IPB), it is important to analyze country and urban data sources. Key information will include demographic data such as the age distribution of the population, what languages are spoken, and what religions are practiced. Again, these factors need to be incorporated into the planning process.

Weather and environmental factors will be the most immediate threats upon deployment to a particular area. Heat or cold injuries can be one of the first threats to be encountered. For example, on my first deployment to the Sinai, we were the first group to jump in and had to do a march in full combat gear to the south camp. As a result, we experienced a number of heat casualties that perhaps could have been prevented.

In addition, it is critical to understand the particular diseases prevalent in the city. Different threats will behave differently in urban areas due to the fact that high concentrations of people can increase the probability of the spread of infectious diseases. Further, if a city's infrastructure has been compromised, then the force may be dealing with contaminated water, raw sewage, or other problems, all of which can lead to epidemic situations. In particular, one should do a close study of high-risk diseases such as HIV, cholera, tuberculosis, dengue, and malaria. These diseases are of greatest concern in tropical and underdeveloped countries. Further, there may be other threats to consider, such as terrorists or the potential for deployment of a chemical or biological weapon.

Before deploying, it also is important to have a good working understanding of the urban medical infrastructure. What is the quality and availability of professional medical personnel, medical equipment, clinics, hospitals, and organization of the health care system in that country or city? Are the NGO activities coordinated? Is there a good medical supply infrastructure? For example, the medical infrastructure may be intact but there may be a shortage of medical personnel, or the distribution network for medical supplies may have been disrupted. Again, AFMIC is a good source of information on a country's medical infrastructure.

Pre-Deployment

- Orders... Set by the CINC
  - Medical Rules of Engagement
    - General Order #1
  - Preventive Medicine Measures
    - Vaccinations
    - Protective Measures
  - Evac Policy and Procedures

During the predeployment planning phase, key considerations include the medical rules of engagement (MROE) and how to ensure that the force understands what to do when assisting with casualties, in particular civilian casualties. For example, in Haiti, the MROE were part of General Order #1. Every soldier was instructed on how to respond to casualties and where to take individuals, whether civilian or military. The U.S. military hospital in Haiti was small, which meant that we had limited capability and resources to take care of patients. Supporting the force was our primary mission. So in Haiti, the MROE dictated that if a civilian casualty was encountered, the soldier should provide first aid and then transport the person to a civilian hospital unless the injury was the direct result of a military action. In general, if the MROE are not clear, then force members will bring all civilian casualties to the nearest military treatment facility, regardless of the patient's condition, how the injury occurred, or the availability of local medical resources.

Key to force protection are the preventive measures required for a given area of operations. Some of these measures will include immunizations, prophylactic medications, safe water, food, and hygiene measures. Not only U.S. forces, but all of the forces in a coalition operation will need to be trained on preventive medicine measures. Often our forces rely on bottled water and plastic latrines while in theater. These amenities are nice to have, but they are not critical and pose a serious burden to the logistical infrastructure. If planning to use other systems, it will be necessary to train personnel on field sanitation techniques and to bring in the right equipment. The regions of the world to which we are now deploying often have limited sanitary infrastructure, so it will be essential to be prepared.

Finally, prior to deployment it is important to determine the evacuation policy and procedures for the force and the civilian population. If not established beforehand, evacuation policies may evolve on an ad hoc basis and potentially become a major cause of friction during the course of an operation.

Pre-Deployment

- Reconnaissance
  - Team: Log, Ops, Epidemiologist
  - Country Survey:
    - POCs: Embassy, USAID, WHO/PAHO, MOH, NGOs, other agencies like CDC
    - Standards of Care
    - Medical Infrastructure

If there is an opportunity to conduct a reconnaissance, then the team should be tailored to include a medical operations officer, a medical logistical officer, and an epidemiologist. Their job will be to conduct a survey of the resources available in-country. Sources of information will include USAID, Ministry of Health, WHO, and other key health care players in-country. The end product of the visit should consist of a good understanding of the medical infrastructure, the prevalent medical conditions, the urban medical needs, and the medical organization present in the city (including that of governmental organizations and of the NGOs).

Questions to address as part of that assessment include: (1) Has the city's medical and other critical infrastructure been overwhelmed? (2) Does the host nation need technical assistance to help reestablish a functioning health care system or just a soft push to provide care for its population? (3) What are the city's or country's medical standards of care? For example, many European countries have standards of care similar to those in the United States, so we may be able to rely on a German hospital to care for U.S. soldiers if necessary.

But in other countries, such as Haiti, we would not be able to rely on any host nation medical assets.

The availability of medical supplies and distribution systems are also of critical concern. Surprisingly, there may often be multiple medical assets already present in an area of operations. For example, in Guatemala City there is a U.S. Center for Disease Control (CDC) field office. In various countries there are at times U.S. military medical laboratories, such as the Navy's laboratory in Cairo, Egypt. The assessment process should identify whether assets like these are present in the theater of operations.

Pre-Deployment

- Mission Analysis
  - Tailor the Package
    - Personnel/Equipment/Supplies
    - Limiting Factors: Transportation & Time
  - Consider:
    - Epidemiological System
    - CSC/Log Infrastructure/Commo
    - Translators

Once the medical command has gathered all available pertinent information, one needs to conduct medical mission analysis. Particular attention should be given to tailoring the medical personnel, equipment, and supply package in order to accomplish the mission. If civilian care is expected, then pediatric, women's health, and geriatric needs ought to be considered in structuring the medical force. I subscribe to the premise that, in general, it is better for the local medical infrastructure and host nation to provide care for its civilians, whereas the role of the military should be to support the local infrastructure with medical supplies, equipment, and perhaps some professional services. However, the guiding rule should be that the military is in a supporting role and that the civilian medical authorities are in charge.

From a planning perspective, key limiting factors will be time and transportation resources. It is essential to have a good understanding of these two factors early on in one's mission analysis. For example, how much time do you have to respond, and what transportation assets will be available? Other factors to consider include the need for combat stress control assets (don't leave home without them), the logistical infrastructure, and the communications infrastructure.

Finally, translators are critical to the medical mission. Even when one may not be planning to provide care for civilian casualties, the Geneva Convention requires that military forces render aid to all patients with true medical emergencies when we come in contact with them. This will require us to be able to communicate with the patient and understand any cultural restrictions so as to be able to provide care and effect their transfer to the civilian medical authorities, if necessary.

Pre-Deployment

- Task Force Organization
  - Understand capabilities and composition of Joint and Combined Forces
    - What other medical assets are in the TF?
    - Assessment of their standards of care
    - Medical issues from home station
    - Preventive medicine status

When in support of a joint task force, one needs to understand what medical assets will be available from each service and how they are to be employed. For example, an Army medical company is not the same thing as a Marine Corps medical company. For joint missions, therefore, the medical planner needs to understand service differences in order to tailor the package appropriately. In the event of a multinational force, such as a NATO or UN-led force, a clinical assessment of each country's capabilities and standards of care is required.

Preventive medicine measures are also critical to mission success. Units must be well trained, disciplined, and resourced to execute a sound preventive medicine program. If there is no soap, how can you clean the dishes? If the cooks are not trained, how can you prevent food-borne diseases?

In addition, there may be particular medical concerns from home station such as chloroquine-resistant malaria, drug-resistant tuberculosis, hepatitis E, HIV infection, and other serious diseases. For example, in Haiti we brought in troops from the Middle East who happened to have a strain of malaria endemic to their country that is resistant to chloroquine. Haiti, on the other hand, does not have a strain of malaria that is drug resistant. In Asia, hepatitis E is quite common, but it is rare in the Americas. Such diseases, if brought into the theater by a multinational force, may tax the theater's medical system and host nation capabilities as well as being politically sensitive. Further, they may affect not only the local populace but also the force. A military force should not introduce a new medical problem into a region that may already be in crisis.

Pre-Deployment

- Mission Analysis
  - Exit Plan
    - What is mission accomplishment?
    - How do we get there?
    - What do we agree to do from the medical stand point?
    - How do we measure success?

Before undertaking operations in urban terrain, one needs to establish an exit plan. This is probably the most important analytical step before executing the mission.

How does one define mission success from a medical standpoint? Each mission will be different. For example, are there specific medical or humanitarian assistance projects to be accomplished or timelines to be met? Will success be defined as when a host nation's medical infrastructure becomes "functional?" If so, what is meant by "functional" will need to be clearly defined. If epidemiological data are to be utilized to measure when mission success has been achieved, then one needs a plan for collecting supporting data.

Ideally, the military will get host nation officials to agree with the exit plan and desired end state as well as how "medical success" will be measured. This facilitates departure, clarifies expectations, and can help ensure a smooth transition from the military to civilian authorities. For example, after Hurricane Mitch I met with the Minister of Health and we agreed in writing what U.S. military medical support would be most useful during the three months of our deployment. It was clear what our exit strategy was. Once these questions are answered, one can proceed with the analysis of how to get there.

Pre-Deployment

- Unit Training
  - Force Protection
    - Non-Lethal Weapons/Measures
    - Convoys
    - Barrier Protection (ID)
  - Language and Culture Training

Two items of force protection that merit special training attention are the use of nonlethal weapons such as pepper spray and convoy operations in crowded urban centers—the most difficult operation in an urban environment from the medical commander's perspective. In terms of nonlethal weapons, it is essential that medics be trained as well as the rest of the force in the use and deployment of these weapons because they are not part of the TTP process.

The entire force must be trained on how to use barrier protection (i.e., latex gloves) when rendering care to any person (military or civilian) in order to prevent the spread of such contagious diseases as HIV infection and hepatitis.

Another training issue that merits special attention is language and cultural training. For commanders, it is essential to have linguists available who have not only have mastered the language but can also do cultural translations. To illustrate, while in Central America I was in a meeting with another senior military officer and the political officer of the embassy. The host nation military officer was being matter-of-fact and quickly moving through the different points of negotiation in order to reach an agreement. However, in the cultural context this style of negotiation was out of character and reflected problems with the relationship that hadn't been picked up by the U.S. senior military officer. Further inquiries discovered that the host nation senior officer felt that he was out of the loop and was being slighted by us, and for that reason he was angry. Once we identified this issue, we were able to work out our communication problems. My point is that it is not enough to be able to converse in another language. You also need someone who can understand the cultural context and provide guidance to you on such issues.

Deployment

- Standard Operating Procedures
  - Civilian Care/ Evacuation Procedures
    - Local coordination
    - Training of the force
  - Casevac/Medevac/Repatriation Procedures
    - US/UN/Civilian
  - Establish a Medical Surveillance System

The care of civilians and evacuation procedures require detailed planning and coordination. It is critical to establish standing operating procedures (SOPs) so that all of the force involved will be able to execute them. When the mission involves a multinational force, the rules that apply to American forces may not necessarily apply to the UN or to the care and evacuation of civilian casualties. Coordination between these entities needs to occur beforehand and not when the crisis occurs—the least opportune time.

If you have to send patients across a national border, it is important to think and plan on how to get the proper visas and documents. In Haiti, when any force soldier was seriously injured we transferred him to Miami. For this reason, we worked out a routine system to get expedited visas at the U.S. Embassy in Haiti for any injured non-American soldier. It is better to plan ahead than have to do so in the heat of battle.

Establishment of a medical epidemiological surveillance system is also critical in order to be able to assess the health of the force and to detect unusual occurrences that may herald preventive medicine problems or the deployment of biological weapons. Such a system enables us to track diseases within a country. Medical surveillance, however, requires cooperation from all the units in the deployed force. The system relies on all units reporting occurrences of such diseases as upper respiratory infections, flu, and diarrhea. This information can in turn help us to determine whether a disease occurrence, such as a case of cholera or anthrax, is within the normal range expected for that country or perhaps represents a disease outbreak or may even signal the possible deployment of a biological weapon.

Deployment

- Mission Execution
  - Care of Force
    - US Standard of Care
      - Supplies
  - Civilian Care
    - Based on their Standard of Care, we are in support
    - Force Protection Concerns
    - Infrastructure Development
  - Exit Plan

It is essential to assure that U.S. standards of care prevail for the whole task force while conducting operations in urban settings, otherwise one may risk division within the multinational force. Ensuring U.S. standards of care can create some interesting dilemmas, such as the use of only FDA-approved medical supplies. For example, U.S. soldiers can only receive FDA-approved drugs, whereas a Turkish soldier may receive Turkish medicine. However, in a coalition operation you cannot have two different standards of care being applied to the force.

At the same time, these standards may not necessarily apply to the host nation. The standard of care provided to civilians should be that of the community and may not necessarily conform to Western-style medicine or standards. In providing care to civilians there should be an emphasis on basic preventive medicine measures and primary care.

The best way to avoid ethical dilemmas is by keeping the local medical civilian authorities in charge of civilian health care. As has been noted, the military's medical role will be one of supporting the civilian medical authorities and infrastructure. Further, NGOs may want the medical mission of providing care to civilians and may be concerned that the U.S. military may supplant them. The best scenario is where the civilian medical infrastructure remains viable and our role is to provide them with a minimal amount of equipment and medical supply support.

In the event of having to support displaced civilian camps, the most important elements will be basic preventive medicine, water treatment, food, vector control, immunizations, black water disposal, and mental health care. I'm convinced that the best alternative is for the civilians to be organized and resourced to care for themselves, with force oversight. Again, do your best to find other organizations like the host government, the International Red Cross, or reputable NGOs to run these camps, if at all possible. The military may provide assistance such as medical supplies or equipment to them. However, the more self-reliant the civilian population is, the easier the transition from military to civilian support will be. Always keep your eyes on the exit plan!

Deployment

- Resources
    - US Military
    - USAID
    - WHO/PAHO
    - NGOs
    - Host Government

With respect to humanitarian assistance or crisis situations such as occur during a major natural disaster, the military hopefully will not be the only "medical top dog" in town trying to assist the populace. One needs to determine what the available resources are locally and assess whether other organizations may be able (or be more appropriate) to assume a particular mission. Government organizations and NGOs bring medical personnel, money, equipment, and supplies to the table; the military should involve them when possible. On the other hand, the military brings to the table uniquely organizational skills that often may not be present in the civilian organizations, especially during a crisis. Some civilian organizations may be reluctant to coordinate with the military, however. Because the military has logistical, intelligence, security, and transportation assets that these other organizations do not have, this can be useful in helping to leverage their cooperation.

Deployment

- Challenges
  - CMOC Coordination
    - Conflict when the main mission is humanitarian
  - Politics
    - US Mission, Host Nation, UN, US Military

There are multiple challenges during a deployment. An important one is avoiding the conflict that can arise during routine establishment of the civil military operations cell (CMOC) or a humanitarian operations cell (HOC). If done improperly, this can lead to a duplication of effort between task force senior leaders and CMOC/HOC personnel. The responsibilities of the Task Force Surgeon or Engineer, for example, and those of the CMOC/HOC personnel, need to be defined early to avoid possible confusion.

Organizational issues and politics will always present a challenge. One needs to understand them as well as possible and work with them. Good medical intentions may have grave political consequences, as the Germans found out in Cambodia. I was told that the German hospital went out of its way to provide excellent medical care to the locals in its sector. The problem was that only one political faction lived in the German sector, and other factions saw this as political discrimination against them. As a result, this good German deed almost broke the political agreement.

328 The City's Many Faces

Deployment

- Challenges
  - VIP Care
  - Contingency Plans
    - Hurricanes, Fire, MASCAL, etc.
  - Telemedicine Employment
  - Keep busy

We discussed providing medical support to the deployed force and to civilians. Another patient population to consider are VIPs such as congressmen, general officers, and U.S. and foreign ambassadors, among others. They will be part of any operation, and one had better be ready to provide care to them and have a plan for doing so. The plan should include how to provide security for them as well as how to provide for their clinical and personal needs.

Contingency planning needs to be ongoing. How will you deal with mass casualties? Are you using nonstandard evacuation procedures? Have you exercised these procedures and plans? How are you going to secure your personnel and equipment in the event of a hurricane, fire, or other natural disaster? How will you sustain your medical operations under different scenarios?

Telemedicine allows us to bring state-of-the-art care to remote areas. We can bring the best minds to bear, especially in cases of clinical or operational matters, using this asset.

The final challenge I would like to address deals with boredom. Soldiers must stay busy, and it is up to us as leaders to find them good employment. Training is always a source of good employment, but it has to be planned and resourced well if it is to be effective.

Re-Deployment

- Execution of exit plan
  - Downsize
    - Personnel/equipment
    - Disposition of supplies
    - Phase out of the missions
- Prepare the personnel for re-deployment
  - Returning briefing
    - Reunion with family members
    - Debriefing

As the force gets ready to redeploy, it is important to do so according to the exit plan. As the operation matures and the desired end state is achieved, the medical effort will decrease and there will be a phase-out of the missions. Expendable medical supplies will most likely remain in country to be donated to the host nation or NGOs, whereas the personnel and equipment will return home.

Before the force is redeployed, it is very important to brief its personnel about impending family reunion issues, and, if necessary, they should receive critical incident debriefings from psychological personnel before departing for home.

Re-Deployment

- Medical Screen
  - Prophylaxis education
    - Required tests and examinations

Remember that no longer will U.S. forces deploy to different regions of the world without having a sound medical surveillance system in place. This is now part of standard procedures. Such a system will enable us to monitor and study exposure to potential risks, and to anticipate and prevent such problems as Gulf War Syndrome. So it will be important to plan for implementing such systems.

The planning and the execution of medical operations in support of urban operations requires a lot of thought and attention to detail. In addition, medical operations have direct operational and strategic implications. For these reasons, they have to be included in the initial mission planning and not simply as an afterthought.

DOMESTIC MOUT PANEL PRESENTATIONS

ANNEX 1: THE ACTIVE ARMY COMPONENT
AND THE 1992 LOS ANGELES RIOTS
BG Edward T. Buckley, USA

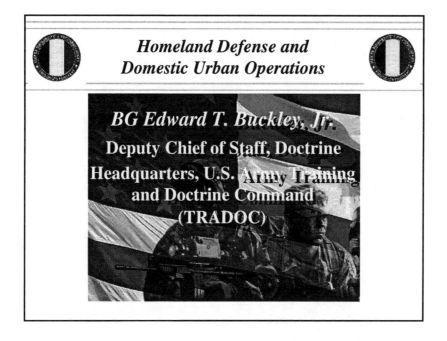

Good morning. I'm very pleased to have this opportunity to talk to you about some of the work we have been doing at TRADOC in the area of homeland defense. By its very nature, homeland defense will almost always include some form of domestic urban operations. We need to make sure that we have the necessary doctrine, training, and equipment to handle them. During the next few minutes, I also hope to be able to share some of my personal experiences as an active component brigade commander during the Los Angeles riots. Hopefully, my comments will stimulate your interest in the subject and lead to some challenging questions during the panel discussions.

Homeland Defense -
A Draft Working Definition

Protecting our territory, population, and critical infrastructure at home by:

• Deterring and defending against foreign and domestic threats

• Supporting civil authorities in crisis and consequence management

• Helping to ensure the availability, integrity, survivability, and adequacy of critical national assets

> **Homeland Defense activities take place in a joint, interagency, and multi-jurisdictional environment**

As most of you know, homeland defense is a new area of emphasis for many of us. Currently, the Department of Defense has no official definition of the term "homeland defense." Therefore, one of the great challenges for us at TRADOC has been to arrive at a definition that accurately captures the multiple dimensions of the task. This is the definition that we have included in our TRADOC White Paper entitled "Supporting Homeland Defense." We believe it properly emphasizes that the Army plays a supporting role to efforts by U.S. civil authorities to defend against attacks on our national infrastructure and other critical assets, whether they come from foreign or domestic sources. We are also working on crafting the appropriate policies, doctrine, tactics, techniques, and procedures for the Army's role in this effort. While we now think of most of our military operations as joint and interagency in nature, the multi-jurisdictional dimension of homeland defense has added another layer of difficulty to the problem. In homeland defense operations, we must work with federal, state, and local government agencies that in some situations do not habitually work well with one another. Even within layers of the government there can be problems, as the well-publicized feud

between the mayor of Los Angeles and the Chief of Police demonstrated to us in 1992.

Categories of Threats

Categories of Threats

- International terrorism
- Domestic terrorism
- Conventional attacks
- Transnational threats

Large civil disturbances may also occur before, during, or after a consequence management operation.

The four traditional categories of threats against which we must be prepared to conduct homeland defense operations include international terrorism, domestic terrorism, conventional attacks, and transnational threats. This audience needs no explanation of these terms.

One of the additional missions that we in the active component find particularly challenging for a variety of historical reasons is responding to riots and other forms of civil disturbances. While not a traditional homeland defense function, civil disturbances of the kind that rocked Los Angeles in 1992 will continue to be a mission for both active and reserve component military forces. Weapons of mass destruction and other infrastructure attacks may also provoke panic or general civil disturbances. In the case of such attacks, the military will be called upon to help restore order so that consequence-management activities can take place.

Homeland Defense-- The Urban Realities

- Cities *will* be a primary target for WMD attacks
- Critical infrastructure *will* be targeted in future conflicts
- Military forces *will* be required to support civil authorities

We must be ready now

It is hard to talk about homeland defense without emphasizing the threat posed to our urban areas by individuals or organizations that have access to weapons of mass destruction. The unfortunate reality is that a WMD-equipped terrorist will most likely attempt to attack a city. The combination of people and media access in cities make them lucrative targets.

Future enemy forces can also be expected to attack cities in an effort to disrupt our ability to deploy our forces to overseas areas of operations. Major port and rail facilities tend to be located in or around urban areas, and defending them will require forces well trained in urban operations.

However, in both situations, WMD attacks and infrastructure sabotage, the requirements for local defense and immediate response will likely well exceed the traditional capabilities of local civil authorities. Therefore, U.S. military forces must be prepared to support the civil authorities across a wide range of possible mission areas.

The Current Response Force

This chart outlines the nature of the organizations involved in typical homeland defense missions and suggests the order in which each organization will likely respond to an incident. While the military must be ready to support civil authorities during homeland defense operations, rarely will the Army be the first unit on the ground. However, rapid escalation of a dangerous WMD situation or a large civil disturbance will require the almost simultaneous deployment of local, state, and federal assets.

Domestic Urban Operations: Likely Objectives and Challenges

• **Provide a visible military presence to help reestablish stability**

• **Prevent and help control:**
   Rioting, looting, fires, and panic

• **Maintain control of "cleared areas"**

• **Impose curfews, if necessary**

• **Support Consequence Management efforts**

• **Protect the force**

• **Maintain Situational Awareness**

Simultaneous, distributed, non-linear, small unit, interagency operations

Whether in support of consequence-management activities or large civil disturbances, the military's objectives and challenges will tend to be similar. Highly visible military forces tend to deter criminal activity and have a reassuring effect on local populations. Military forces can also work closely with local law enforcement agencies to help maintain order after the initial response activities. During homeland defense operations, one challenge for the commander will always be to protect his forces without engaging in the application of unnecessary force against American civilians.

An additional problem for the commander will be establishing and maintaining an accurate picture of the urban "battlespace." As we found in Los Angeles, local law enforcement agencies will usually serve as the hub of this effort. They are familiar with the local area and will have detailed maps and information networks already in place.

Finally, from an operational perspective, military operations in American cities will take on many of the same characteristics that we assume for them in other situations. Small units will be conducting localized operations throughout the city in close coordination with law enforcement officials, individuals from federal agencies, and a wide variety of additional "partners."

As we all know too well, urban warfare is hard in the best of situations. Even when we are operating literally in our own backyard, as is the case in domestic urban operations, situational awareness becomes the key element necessary for our successful accomplishment of the mission. The Los Angeles riots, the Oklahoma City bombing, the World Trade Center bombing, and dozens of smaller similar events have taught us this lesson again and again.

Domestic Urban Operations:
Ongoing Issues

Command and Control

- **Unity of Command**
  (Federal, State, Military)

- **Multiple levels of approval**

- **Centralized planning (ROE, tasks)/Decentralized execution**

- **State/Federal status of National Guard units**

- **Compatibility/Integration of communications/intel systems
  (military, government agencies, civilian law enforcement
  agencies, cellular phones, fax, e-mail, police scanners,
  messengers, …)**

There are a number of issues we are currently examining with respect to urban homeland defense operations. They tend to fall into two large categories: command and control issues and operational concept issues.

We have already mentioned most of the command and control issues. Military forces will usually be employed in support of some civil authority—local, state, or federal. We need to examine how we will communicate with these civil authorities and develop processes to ensure that our doctrines and policies are compatible.

Because of the sensitivity of domestic urban operations of all types, centralized planning and decentralized execution will be the order of the day. Small-unit operations are by their very nature decentralized to a certain extent, and this will certainly be the case in domestic urban operations.

One command and control challenge that must be recognized by active component military planners is that the status of National Guard assets may change over the duration of the operation. They may be under the governor's control as the mission begins, become federalized during part of the operation, and then return to state control to better facilitate law enforcement activities.

One final command and control challenge is a technical one. The wide variety of communications platforms in use in domestic urban operations will tax the communications SOPs in even the best organizations. However, each of the civil systems in use contributes unique capabilities to the commander's situational awareness, and so each must be incorporated as fully as possible.

Domestic Urban Operations: Ongoing Issues

Operational Concepts

• **Training to support GARDEN PLOT OPLAN**

• **FBI is the lead for federal training support to local efforts**

• **Active/Reserve component training for domestic support**

• **Joint military doctrine for domestic urban operations**

• **Small unit operations versus "riot control techniques"**

The second large basket of issues contains challenges of an operational nature. Most active component units do not have the resources to adequately train for the tasks outlined in domestic support planning documents. Add to this challenge the fact that federal agencies other than DoD have the lead responsibility for training local law enforcement agencies to handle domestic incidents, and we can end up with some serious operational concept mismatches. We need to make sure that both our active and reserve component doctrine and training support the type of homeland defense or civil disturbance missions likely to occur. One lesson we took away from Los Angeles is that small units need to be able to employ both riot-control techniques and more aggressive small-unit tactics as the situation on the ground changes.

Homeland Defense: An Urban Challenge

Khobar Towers

King Abdul Aziz Air Base

Dhahran, Saudi Arabia

The Pentagon

Washington, D.C.

Just as they long have our forces overseas, terrorists and enemy military forces are now targeting the military forces, civilian population, and critical infrastructure in American cities for destruction. It is our responsibility to protect them. And while MOUT takes on some unique dimensions when we talk about it in the domestic context, many of the concepts and skills we are currently working on developing will prove invaluable when we have to conduct urban operations in the streets of America.

Thank you for your attention. I look forward to your questions.

ANNEX 2: FBI CONCERNS DURING DOMESTIC MILITARY URBAN OPERATIONS
Mr. Jim Rice, FBI

Washington Field Office Talking Points

1. The WFO has made response to WMD incidents a priority.

    •   The largest increase of agent assignments is in counter-terrorism squads.

2. The FBI has initiated several plans to combat this threat:

    •   Creation of a WMD coordinator in each office; the WFO has the largest contingent of WMD personnel.

    •   Creation of WMD teams, similar to ERT teams; the WFO has the first team in FBI.

    •   Enhanced bomb tech capabilities; the bomb data center now has the most advanced anti-bomb capabilities in America, including DoD assets.

    •   Creation of HMRU and increased lab capabilities.

3. The FBI, at the request of the Attorney General, has created the nationwide National Domestic Preparedness Office (NDPO) to assist state and local authorities with major terrorist incidents and WMD preparation.

4. The National Infrastructure Protection Center (NIPC) has been established at FBI Headquarters.

    •   The center exists as an international center/joint agency effort to counter threats to U.S. infrastructure systems, including cyber-terrorism.

5. Training for agents has increased awareness of the WMD threat and the nature of the threat:

    •   The threat from military-style weapons versus improvised devices and pathogens.

- While military nerve agents are deadly beyond belief, improvised devices are not usually similar in lethality. Homemade pathogens, poisons, and chemical compounds, while deadly, have not had sophisticated delivery systems. This means the chemical may be deadly, but there is no good way to deliver it, meaning mass casualties are virtually impossible to achieve. The anthrax letters are a perfect example. Even if the letters had been real, only the person who handled the plastic bag inside the letter, opening it up and breathing in or eating the spores, would have been affected—not the entire building. A homemade pipe bomb is probably a threat to a greater number of people than a homemade chemical/biological agent.

6. WFO/National Capitol Region specific FBI responses to threat:

- Addition of WMD-trained agents over the last year (500 percent increase).

- Creation of first FBI field office WMD team.

- Addition of a chem/bio capability to the WFO ERT team.

- Addition of chem/bio capability to WFO SWAT team.

- Formation of the "National Capitol Special Response Unit" to co-locate and coordinate all WFO ERT/WMD/SWAT/bomb tech capabilities and efforts, including rapid deployment.

Coordination with Other Agencies

1. The WFO is dedicated to the idea that all WMD responses must be a group effort.

2. The FBI is the lead agency for crisis management under PDD-39, but it leads a team; this is not a single-agency effort.

3. State and local authorities will not relinquish their vital role as first responders in these events.

4. Local/state authorities will always have an equal voice in the decisionmaking process:

- The role of the FBI is unchanged; it is the lead crisis management/investigative agency, not the lead political entity. We are a law enforcement agency. That's our role.

5. In the National Capitol Region, the FBI will function as first responders also. Increasing threat, specifically to the federal government and the continuity of government provisions, requires this function. However, everything is to be done in conjunction/coordination with local authorities.

6. There is continued joint training/exercising/intelligence sharing with all other agencies, including specific DOD assets.

7. There is an increase in the size of joint terrorism task forces

8. A terrorism working group has been created in Washington, D.C. for non-JTTF agencies.

9. DoD/national guard assets are included in contingency plans

- The activation of DoD assets requires either governors' activation of national guard assets or the Attorney General requesting DoD assistance via national command authority; this is unchanged. However, the process has been streamlined for speed and clarity of authority. Civilian authority is not relinquished when this occurs: DoD assets simply assist local/federal authorities at the scene with their increased numbers or specialized skills and materials.

10. The transition from crisis management (FBI) to consequence management (FEMA) is being practiced and streamlined.

- The federal response plan is constantly updated

11. The FBI Strategic Information Operations Center (SIOC) is the national-level command post expanded to handle several Oklahoma City–sized incidents at the same time. It is the most sophisticated civilian command and control structure in law enforcement today. It can and will function as a joint operations center (JOC) for the national capital.

Washington Field Office Response Mission

The FBI, as the largest federal law enforcement agency, was given the leading role in responding to terrorist incidents in the United States under Presidential Decision 39 (PDD-39). Also by presidential decision, the FBI is the lead federal agency dedicated to responding to any use or threatened use of a weapon of mass destruction for crisis management. This, however, does not mean that when the FBI arrives on the scene of an incident it pushes all other agencies aside. Rather, it requires the FBI to identify, support, and work with any and all other agencies involved due to their mission or jurisdictional requirements. The FBI takes this role seriously and has dedicated itself and its resources to increasing our ability to meet this challenge now and in the 21st century. The Attorney General of the United States agrees with these mandates and supports all the efforts of the FBI in this regard.

The federal government's role in consequence management is delegated to the Federal Emergency Management Agency (FEMA). FEMA's role in these events is to support local and state law enforcement and emergency service agencies. A major portion of the responsibility for law enforcement activities during an incident like this lies within the sphere of state and local authorities. In the event that a governor deems that these resources are overwhelmed by either the scope of the disaster or the length of time of the recovery, the governor can activate his National Guard and request other federal assistance, such as that from the Department of Defense or Public Health Service. Federal policy requires that all such requests during the crisis phase of an operation be submitted via the local FBI office. The local office then forwards the request to FBI Headquarters, where these other federal assets already have representatives prepositioned. However, now as always, it has been the policy of the FBI to respond to any request for assistance in an emergency to prevent the senseless loss of human life.

Of major concern to everyone in law enforcement during the last decade has been the rise in the specter of terrorism. The Washington field office now has dedicated more resources to counter this threat than at any other time in the history of law enforcement. The creation of new methods, technologies, and abilities to counter this threat gives the FBI the ability to stay one step ahead of these threats. Increased funding has given the FBI the ability to respond to major acts of terrorism anywhere in the world that an American is threatened. Increased abilities at home have given each FBI field office the ability to investigate, coordinate, and counter the threat of chemical or biological terrorism as never before. Increased cooperation among law enforcement and intelligence agencies has given the law enforcement community the ability to detect and deter acts of terrorism prior to the event rather than simply responding to it after the fact. Preventing another Oklahoma City is certainly preferable than responding to, investigating, and solving another terrorist incident

These abilities were recently seen in the Washington field office's response to the bombing of two American embassies in Africa. The worst-case scenario, two major acts of terrorism in two different countries far from the supply lines of the United States, finally occurred. The WFO responded by sending teams to both sites within hours to evaluate the scope of the situation. FBI legal attachés, stationed in American embassies across the world, responded first, followed by FBI agents from Washington, D.C. Immediately, FBI agents with special skills, such as evidence recovery teams, explosive technicians, terrorism experts, investigators, communications specialists, and intelligence analysts were dispatched by both commercial and military airlift to the scene.

This constituted the largest overseas deployment of FBI personnel since the Second World War. During the African deployment, teams of FBI agents, in conjunction with the local African authorities and in coordination with the Department of State, conducted extensive crime scene searches, recovered and cataloged a tremendous amount of evidence, accumulated intelligence on the event, and facilitated the identification and arrest of several of the terrorists involved in the event. The other terrorists, who managed to slip through an ever-tightening net of international law enforcement, are today among the most hunted criminals in history. The WFO still has agents in Africa today, continuing to hunt for these criminals

At home, the Washington field office's ability to respond to the ever-increasing threat of the use of a weapon of mass destruction, such as a chemical or biological weapon, has vastly increased. Each FBI field office, regardless of size, now has a weapons of mass destruction

(WMD) coordinator to facilitate WMD investigations. WFO has five agents assigned full time to this task—the largest number and the best trained of any FBI field office. These experts make their knowledge, expertise, abilities, and network of other chem/bio experts available to state and local law enforcement whenever needed. These agents can help coordinate the activities of the FBI with these local law enforcement agencies and fire departments, HAZMAT teams, and they also increase the intelligence available to these units, something that was virtually unknown two years ago. Law enforcement has traditionally been reluctant to share information with non-law-enforcement agencies such as the fire departments. Now the WFO has dedicated resources to do just that.

The FBI has also created a national response team for major chem/bio incidents: the hazardous materials response unit or HMRU. This specialized unit, which is a division of the FBI laboratory, has the ability, much of it classified, to detect, deter, and mitigate WMD events anywhere in the United States. The bomb data center of the FBI, the national center for training, research, and development, and coordination for civilian bomb squads have all greatly increased the FBI's ability to respond to and handle large and sophisticated explosive devices in the United States. In addition, the FBI, at the request of the Attorney General, has formed The National Domestic Preparedness Office (NDPO). This office exists for the single purpose of helping state and local agencies to acquire training, resources, information, and increased abilities to respond to incidents of terrorism. All of these national assets are located within the National Capitol Region and the WFO division; they work closely with WFO agents to provide support directly to the area.

The Washington field office of the FBI, the field office with the direct responsibility for the District of Columbia and Northern Virginia—the National Capitol Region, has greatly increased its ability to respond to terrorism and WMD incidents. The Evidence Recovery Team in the WFO is the most experienced in the FBI today. The bomb technician complement is being greatly enhanced (400 percent) at the Washington field office, and today coordination and cooperation between area bomb squads is at an all-time high. The WFO's weapons of mass destruction program is the largest and most sophisticated of any FBI field office. The team, which includes agents with advanced degrees in nuclear engineering, chemistry, and hazardous material management, also obtains specialized training from the Department of Defense, FEMA, other federal agencies, and the private sector. In addition, for the first time, WFO has dedicated resources to coordinate WMD activities directly with the HAZMAT teams of area fire departments. As these teams are often the first called to a chemical incident of any type; this is an essential link in the chain, but an often-overlooked link until this time. Also, the Washington field office is reorganizing its existing resources to facilitate a faster, better-equipped, and more streamlined response to these incidents. This restructured unit, a "National Capitol Region–Special Response Squad," will have the ability to handle a multitude of incident types within the region.

The WFO's dedication to counter the specter of terrorism, both in the nation's capital and abroad, is ever expanding. The FBI currently has 16 joint terrorism task forces across the United States, of which the WFO's is the second largest. These joint agency counterterrorism units have proved so effective at their mission that the FBI is going to fund the expansion of these units to every major metropolitan area in the United States. Plans for an expansion to almost 40 of the task forces are planned. The FBI's dedication to this effort is based on long-range plans for the FBI to deal with this threat for decades to come. The WFO's dedication to constantly prepare to face this threat will not falter, but grow with the cooperation with all of the partners in this fight.

ANNEX 3: WMD IN URBAN OPERATIONS
COL Daniel F. Uyesugi, USA

Weapons of Mass Destruction in a Domestic MOUT Environment
COL Dan Uyesugi

The purpose of the briefing today is to discuss weapons of mass destruction in an urban environment. You've heard about the riot control and the civil disturbance parts. You've heard of the terrorist part. Well, there are also threats from weapons of mass destruction; they create a unique challenge for response forces and response planning.

"The proliferation of Weapons of Mass Destruction and the means of delivering them continue to pose an unusual and extraordinary threat to the national security, foreign policy, and economy of the United States . . ."

-President Clinton

As General Buckley indicated, two significant strategic changes have taken place since our overwhelming victory in Operation Desert Storm. The first was a push by many nations and terrorist organizations to acquire weapons of mass destruction and the means to deliver them to help compensate for the West's conventional asymmetric advantage. Secondly, we have created enclaves here in the United States for our military forces. And what we've done by becoming more of a CONUS-based military force and relying on a very select number of power projection platforms is that we have created some very visible targets of opportunity for weapons of mass destruction. They're very convenient targets and they're very vulnerable, and not all of them are military installations. Many of them are civilian airfields. Many of them are ports. Many of them are logistics and commercial communications facilities. And we have to protect all of them. They are national strategic assets and it's part of our mission to protect them.

Weapons of Mass Destruction

- Nuclear
  - Detonation
  - Orphaned Radiation Sources
  - Radiological Dispersion

- Biological
  - Bacteria   - Viruses
  - Rickettsia   - Toxins

- Chemical
  - Nerve   - Blood
  - Blister   - Choking
  - Riot Control

Nuclear

- Detonation: A fission or fusion device explodes. Requires "weapons-grade" fissile material, detailed technical knowledge to build. Russia's internal condition makes it possible to buy: no knowledge required, only cash.

- Radiological dispersion: Radioactive material is dispersed by mechanical means. Does not require weapons-grade fissile material or detailed technical knowledge.

- Orphaned radiation source: Legal sources (licensed to DOT, construction companies, hospitals, etc.).

The slide provides examples of biological and chemical agents.

Weapons of mass destruction are in fact NBC materials, nuclear, biological, and chemical materials, that create hazards. They kill, injure, or incapacitate. They also contaminate material and environments. In many cases you have to remove hazardous materials, you can't destroy them. You have to neutralize some chemical and biological materials, and afterwards you have to prove that you've met civilian safety and occupational health standards. The effects can be tailored. We've always had the nuclear weapons threat, but in the past, it's been a very high-tech business requiring weapons-grade materials and high technology. With the disintegration of the former Soviet state, that technology has become more readily available, both internationally and here in the United States. There are orphaned radiation sources that might be a contamination source. Materials from such sources can be dispersed as radiological dispersion weapons. I'll talk to you about one example of that from Desert Storm below.

Biological materials include toxins and materials that cover the entire spectrum of pathological materials. Toxins are basically poisonous chemicals that are of a biological origin. As for chemicals, they have a wide range of possible effects on human beings. They include attacks on your nerve system, blisters on the skin, various incapacitating mechanisms, choking, and blood agents. These materials can be either persistent or nonpersistent. They may be in vapor, liquid, or solid form.

Threats

- Tokyo Subway
- Narita
- Yokosuka Naval Base
- Matsumoto, Japan
- Oregon Salad Bars
- Moscow Park
- Sydney, Australia

- Tokyo subway, 1995. Killed 12, injured 5,000. Aum Shinrikyo cult (ASC).

- Narita airport, Japan, 1990. ASC attempts anthrax attack using vehicle-mounted aerosol generators. Unsuccessful due to nozzle problems.

- Yokosuka naval base, Japan, 1990. ASC conducts unsuccessful anthrax attack (see Narita above).

- Matsumoto, Japan, 1994. ASC conducts partially successful sarin attack using vehicle-mounted aerosol generator, kills 7.

- Oregon salad bars, 1984. A religious cult sprays salad bars in The Dalles, OR, with salmonella, to attempt to alter local elections by reducing opposition turnout. 750 sickened.

- Moscow park, 1995. Chechen terrorists plant an RDW in a park in Moscow. Authorities locate and remove/deactivate before it functions.

- Sydney Australia, 1996, 1997. A series of chlorine bombs in Sydney suburbs.

There are several great examples of nuclear, biological, and chemical terrorism. The Tokyo subway in 1995: 12 killed, 5,000 injured by a cult. This was a sarin nerve agent attack. There was a similar attack in Japan in 1994 by the same cult. That time they killed seven. There have been several attacks with anthrax. There was also an attack in Oregon in which salmonella was spilled across a salad bar that caused 750 to become sick. This was an attempt by a local religious group to affect the outcome of an election and keep these 750 people from voting. A Chechen terrorist planted a radiation dispersal weapon in a Moscow park in 1995. The local authorities had to go in and render that device safe.

What history is telling us is that terrorist threats are real. They can kill and injure. They can certainly terrorize. In the United States as well as overseas, they are low tech and include all the full range of nuclear, biological, and chemical threats.

Threats

- Accidents/Incidents
  - Bhopal, India
  - Chernobyl
  - Sverdlosk

- Orphaned Radiation Sources
  - Hospitals
  - Construction Sites

- Flame

Accidents/Incidents

- Bhopal, India, December 3, 1984. 2,000 killed outright, up to 300,000 injured, as many as 8,000 have died since.

- Chernobyl (Chornobyl), Ukraine, April 26, 1986. 31 site workers killed in accident. Between 1986 and 1990, 5,700 cleanup workers died, according to Ukrainian government figures (mostly men in their 20s).

- Sverdlosk. Estimated 1,500–2,000 casualties, including cleanup crews.

- Milwaukee, 1993. Water supply, crypto sporidium, 400,000 casualties, 111 fatalities.

Orphaned Radiation Sources

- Hospitals. Cobalt 60 source found in a dump by a man paid to monitor dumps for low-level radiation. In briefcase, no shielding, lethal dose in under an hour; how did it get there?

- Construction sites. Sources contained in machines that gauge the thickness or density of roads, foundations, pilings, etc.

There are other possible threats that are not overt. These incidents involve nuclear, biological, chemical, or industrial facilities. In Bhopal, India in 1984, 2,000 were killed outright, 300,000 were injured, and many have died since. At the Chernobyl nuclear power plant complex in 1986, 31 site workers were killed in the initial accident, and between 1986 and 1990, 5,700 cleanup workers were reported to have died. We've had a number of reported finds of abandoned cobalt 60 sources found in dumps, abandoned office spaces from hospital radiological analytical equipment in hospitals, or construction sites.

Flame

And, finally, one of my favorite areas is flame. What happens if you throw a concussion or CS grenade into a structure with propane tanks? You're going to blow the whole structure up. If you're working in the DEA environment, drug laboratories often contain combustible solvents. Combustible materials have been also deliberately used as weapons. Commercial propane bottles are a favorite terrorist means of boosting explosives. When Margaret Thatcher was nearly killed in an English hotel, the front three stories of the hotel came down. Two recreational propane bottles and a quarter-pound block of TNT brought the whole front of that building down. Luckily, she was in the loo at the time and survived.

Considerations

- Detection
  - Medical Monitoring

- Spread of Contamination
  - Ventilation Systems

- Enclosed Space
  - Oxygen Displacement
  - Flammable Atmosphere

There are several considerations that I'd like you to think about as we discuss this. Early detection is absolutely essential to support evacuation and the control of casualties. Unfortunately, in the case of nuclear, biological, and chemical exposures, symptoms may be the first indicators. That's where the medical community becomes so important to us. Emergency medical care providers, and certainly security and fire teams, must be aware of the potential terrorist use of NBC materials. They must know how to respond and who to report to. With respect to spread of contamination, urban areas tend to channel NBC effects. You heard from General Delk that riot control agents don't go where you think they're going to go in an urban environment. Not only do cities channelize and contain the NBC effects and hazards, they also conduct them into the ventilation systems. One of the things that happened in the Tokyo subways was that the Tokyo police set up their command post right on top of a subway ventilation system. Guess what? They had casualties associated with the ventilation of the agent up through the subway vents and they had to move their CP, but they lost eight people who were incapacitated. Subways are an ideal place to employ biological weapons. They are out of the sunlight; biological agents don't like the UV rays that are a part of sunlight. It's a moist environment and it has good temperatures. If you move them along the tracks, the agent can be taken from one station to the other. You get a lot individuals going in and a lot of individuals departing. It's an almost ideal place for contaminating a large number of people. What's the number one contamination related problem we have? Casualties. We've got to make sure that we keep the casualties under control and clean them up. We cannot allow them to take contamination into the hospital. Contamination is beyond most hospital capabilities.

Here are some discussion points I'd like you to think about. Attacks must be treated as real until such time as we can absolutely prove differently. Was it a weapon of mass destruction? Is there a residual hazard that's of concern? Where in the heck is it? What is it? How do we determine who is affected, so that we can get them under control and get them support as quickly as possible?. Who must be notified to get help? Where can you get help?

Two final areas of concern:

Enclosed Space

- Oxygen displacement: in enclosed spaces, oxygen can become displaced. Rescue personnel should be equipped with an O2 monitor. Do they need SCBA to accomplish their mission?

- Flammable atmosphere, dusts, aerosols and vapors can create explosive atmospheres. How do you detect this condition, especially if you need to use distraction devices or CS grenades/projectors? And are your flashlights mine-safety approved?

Emergency Response

- Identification
  - First Responders

- Confirmation
  - RAID Team

- Mitigation
  - DOMS

Identification

- First responders: Must recognize that the event is a WMD event.

Confirmation

- RAID Team confirms (see backup slide).

Mitigation

- DOMS makes military units available to support the on-site commander (see backup slide).

We've got to train those first responders, those emergency medical personnel, security force personnel, and fire department personnel to recognize the key indicators of WMD use.

Discussion

Explicit threats have to be treated like attacks until until proven otherwise.

Questions to be answered:

Was it a WMD attack?
Is there a residual hazard?
Where? What?
Who was possibly affected?
Who must be notified?
Who can you ask for help?

COL DANIEL F. UYESUGI

RAID Team Concepts

- Mission
  - Assess
  - Advise
  - Facilitate

- Operational Phases
  - Pre-incident
  - Alert
  - Deploy
  - Response
  - Recovery

- Training
  - Certification
  - Validation

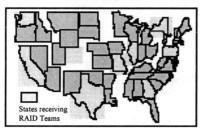

States receiving
RAID Teams

Backup slide 1

NBC in Stability & Support Operations

FM 3-21 under development, publication 4Q FY99

- Supports FM 100-20, *Stability & Support Operations*
- Supports consequence management
- Geared toward Chemical Units and NBC staffs

Chemical Corps Role in stability & support operations

- NEO
- Arms Control
- Support to Domestic Civil Authority
- Humanitarian Assistance and
  Disaster Relief
- Security Assistance
- Nation Assistance
- Support to Counterdrug operations

- Combating Terrorism
- Peacekeeping Operations
- Peace Enforcement
- Show of Force
- Support to Insurgencies &
  Counterinsurgencies
- Attacks and Raids

Backup slide 2

C/B-RRT Mission

Deploy to support LFA with crisis and consequence management under possible control of the CINC/RTF or JSOTF, as directed

- Provide dismantlement (render safe), transport, disposition/disposal, and neutralization support

- Provide agent monitoring, hazard prediction, detection, laboratory analysis, mitigation, and containment support

- Provide medical advice and support for patient decon, triage, transport, and treatment

- Provide advice and/or technical expertise on chem-bio issues to federal, state, and local agencies

Backup slide 3

C/B-RRT Summary

- C/B-RRT is tailored from pre-existing, specialized, joint assets

- C/B-RRT under operational control of CINC **or** specific assets under operational control of LFA

- C/B-RRT Supports crisis and consequence management

- C/B-RRT leverages geographic distribution for faster response

- C/B-RRT deploys an organic command and control system with reach-back capability

Backup slide 4

CBIRF Mission

"When directed, forward deploy Domestically or Overseas to provide Force Protection or Mitigation in the event of a CBR-N incident. Be prepared to initially respond to no-notice CBR-N incidents with a rapidly deployable response force. Continue to be innovative in the development of CM concepts, doctrine, organization, tactics, techniques, procedures and equipment. Conduct force protection training for Fleet Units and Assist Federal State, and Local Response Forces in Developing training programs to manage the consequence of a CBR-N incident."

Backup slide 5

CBIRF Background

- Established in April 1996 as part of CMC's planning guidance

- In response to the increasing threat of a CBR-N terrorist event (PDD-39)

- Available to Marine force Commanders and the National Command Authorities

- Suited for operations in a wide range of military-civilian contingencies

Backup slide 6

ANNEX 4: Q & A FROM DOMESTIC MOUT/WMD PANEL

Question: Would it help to deploy miniature biological and chemical sensors and video cameras in all major cities? My second question is, would the use of foam have helped during the 1992 Los Angeles riots?

MG Delk: I'm going to answer the last part and look for help on the first part. First of all, foam would have been of no help. Potential targets very rarely got close to us. I can remember one case in which a gangbanger did get close. He came up to a soldier and tried to take the rifle out of his hands. We were exercising a lot of restraint, but not so much restraint so that his NCOIC didn't butt stroke the gang-banger. That was the last time they got close to us. As for biosensors and video cameras, I'll look for help from my compatriots here.

BG Buckley: I don't think it's affordable, number one, and I'm not sure our society would allow it, number two. We would also have a hard time affording a good sensor suite just to accommodate the military requirements.

Question: A common theme that I heard each of you discuss dealt with the difficulties in making progress via the interagency process. Do lessons from homeland defense operations offer anything we can turn to and use as a guide?

BG Buckley: I would tell you that there is some energy being generated on this topic. The Department of Justice is the lead agency for many of these activities. The President has talked about it, so I think the opportunity is there, but it's hard to get the various agencies to take time out because they typically, unlike the military, don't have a lot of planning capability. Most of them are in the execution business. Their budgets don't allow for sitting down and working out SOPs or how they are going to cooperate. So even though I'm cautiously optimistic, it's going to be a struggle.

MG Delk: I'd also point out that the Army's JAG school in Charlottesville is going to be getting together to address some of those issues. I know there is Army-wide concern about the issue.

Question: You mentioned intelligence integration problems. What can we do on the military side that would help with the integration of our intelligence-collection effort in homeland defense matters?

MG Delk: First of all, I would tell you that I didn't see it as a problem. Any problems disappeared almost immediately and everyone involved threw their arms around each other and made it all work. Any time you go into something like that there's going to be an intelligence void when the event is not anticipated. Our military intelligence community is very quickly getting a handle on intelligence preparation of the battlefield and getting word to the troops. If there's any part of the Army that has made great strides in the last twenty years, it's our intelligence community.

Mr. Rice: The biggest thing we've seen is that you can't wait until the event happens to meet each other over the smoking hole in the ground. You've got to know who your counterparts in other agencies are beforehand. In my case it's a little easier because I've got ten different agencies and we all sit there and share intelligence. There are agencies in addition to those from the DoD. We sit down once a month.

BG Buckley: The military intelligence community has to be careful to identify where their lane is. In domestic environments, the FBI and local law enforcement agencies are the collectors and processors of information and we're a recipient. Our intel folks can do some analysis and provide information, but we can't do any collecting.

It gets to the command and control issue. You had a couple presentations here today that have addressed the lack of interface between the communications systems. We've got a problem in the military, but ours is very small compared to the problems within communities, between the police department, the fire department, the county, the state, and federal agencies. None of these organizations' systems talk to each other. It's incredible. So when you go into emergency situations there is a need for a whole bank of radios and liaison people just to have the ability to interface. Unfortunately, everybody makes their budget decisions without consulting one another.

COL Uyesugi: Speaking for the combat development side in the U.S. Army Chemical School, I know that we have taken a look at each one of our critical battlefield operating systems and our NBC reconnaissance vehicles to ensure they are Y2K compliant. But I heartily support what General Buckley just said and General Delk's previous remarks. The big problem is lack of interoperability between the various federal agencies. Period. Even assuming that we're Y2K survivable, the C4I systems we have must be compatible. In many cases we're not even compatible on the same military installation, let alone with systems out in the community itself.

Question: As the National Guard and Reserves start to take on the role of homeland defense, what do you think the challenges are?

MG Delk: First of all, homeland defense is not new to the Guard. We've assumed that mission for over 200 years, but our role has been greatly expanding recently. As was briefed, we now have these RAID detachments. They're being given a very complex and important mission. We're starting off with 10 of them. They won't be operational until January 2000. We will have some additional light RAID detachments that will be coming along later, some 44 of them. All of this is expensive; all of it takes time, and even then it isn't going to completely solve the problem.

COL Uyesugi: Sir, if I could pick up on that. This is a real challenge for the total force. We've got to standardize the entire DTLOMs, or doctrine, training, leader, organization, material, and soldier-side, of this. Right now we've got a number of different organizations going out and claiming they know what is needed. Well, it has to be brought under a single organization, and that's what TRADOC does as a combat developer. But for the military forces, when it comes to TO&E and TDA go-to-war units, we have doctrine. We have standardized training based upon an analysis of METL, those basic tasks that are critical for an organization. What we have not done is determine what this civil support mission is all about and how that impacts upon the total Army and its ability to "go to war." Right now we're still tasked to go to two major theaters of war. We're resourced to do that. We're not resourced to do the homeland defense mission yet, and we haven't established a civil support joint task force command that we understand is yet to come. To standardize this, we'll first have to get the mission, which FORSCOM doesn't have quite yet. Second, we've got to deconflict what we're asking active component, Reserve, and National Guard units to do in support of homeland defense and civil support. You have units that are TPFDL'd and have missions to go to a CINC overseas. Now you're going to assign them an additional duty that could get them decisively engaged in a city or a state on four hours' notice within any one of ten federal regions. We're not resourced to do that yet, but we're doing it as an additional duty in our chemical defense, engineer, military police, transportation, and medical units. All of these units are going to be asked to come forward and support regional RAID teams and relevant civil and federal agencies in association with this domestic support mission. You have to remember that National Guard and Reserve units only have about 37 days a year to train. Now you are asking them to add entirely different tasks to their training load. The equipment for such missions is often nonstandard. It's not TDA. In many cases it may be CTA—common table of allowances. Soldiers have not been trained how to maintain that equipment. That's going to have to be done centrally; it is going to be a bit of a challenge. We've got to bring 180 units on board in the TOE Army to support the 10 RAID teams over the next two years. There are some real challenges yet to be addressed.

Question: You mentioned anthrax and other biological agent attacks. What if it's an unwarned attack? How will we get health care providers on the front lines in time? How do we determine that those five people with headaches are the leading edge of an anthrax exposure?

COL Uyesugi: We need to reach out to the health providers, first responders, and first caregivers and give them an appreciation for exactly what a nuclear, biological, or chemical agent might do and what symptoms would look like. That requires specialized training. Now when you get the fifth, sixth, seventh cases, a large group of affected individuals from the same location, the authorities will likely start to realize that a problem exists. Such cases would go through the public health service into the Centers for Disease Control and should result in notice of an attack.

Question: Can you describe the RAID team organization?

COL Uyesugi: The RAID teams' mission is to assess incidents on a regional basis. There is one team for each of the ten standardized federal regions. They're to advise the civil authorities on what type of agents might have been involved, the hazards associated with these agents, and the type of immediate action that might be necessary. The teams would then facilitate any support that the federal government might have available. Initially they're supposed to go through new equipment training. They'll also go through validation training under the direction of the Director of Military Support and then undergo certification before being turned over to Forces Command and the two continental U.S. armies.

BG Buckley: The ten teams are AGR Title 32. They are going through very specialized training at this time. The initial ten teams are regionally located so they have the ability to respond quickly. They'll use Air Guard assets to get to their destination in many cases.

Question: I had a couple of questions, one for Colonel Uyesugi. I was a little bit troubled, I guess, by one statement that you made concerning biological agents in which I understood you to say something to the effect of "We could haul these people off to hospitals, give them antibiotics and we don't really need to worry about them too much because they could be cured by antibiotics." Is there recognition of increasing natural antibiotic resistance worldwide? Of deliberately designed antibiotic-resistant threats? The second part to that question: have you given a lot of thought to agents that can be transmitted person-to-person and the facilities required to contain a contaminated patient population? My last question was for General Delk. You described Guardsmen's twelve-hour work/rest cycle. It sounded like these soldiers were sleeping in unoccupied buildings, and I'm wondering what their exposure to rats, mice, other vectors was.

COL Uyesugi: My degree is in nuclear physics and chemical engineering and I will have to defer to the medical community in a moment. If you can start the antibiotic therapy early enough, you have at least a chance of preventing death. It is my understanding that prior to symptoms manifesting themselves, antibiotic therapy gives you an increased chance of surviving the exposure. Once you've had a vaccination, it gives you that much better a chance of resisting the initial exposure and surviving with the help of antibiotics later. Early notification of exposure is thus important both on the battlefield and in the civilian community. It's important to know who was exposed and to get them under control and observation.

With respect to the agent threat, the licensing of vaccines is an extremely time-consuming, difficult, and costly venture. It's not something that any one nation can take on alone; it's currently being undertaken on a multinational basis with a number of different countries, specifically the Canadians and the Brits. Under JVAC they are in fact pursuing eleven different biological vaccines that are based on an international task force listing of threat agents and variations thereof. Each of the three nations is looking at the research and development side, side effects, and licensure and indemnification requirements. We're going to share in the long term. But that's a long-term plan. Smallpox is a big issue right now. I'd refer to Roy Flowers and the folks from the medical center and school if you want some more information.

MG Delk: To answer your question about hygiene, it was an issue. First of all, the soldiers thought they might be on duty for two or three days. They ended up being there for weeks. They didn't bring enough underwear. But like any good soldier does, these men and women solved the problem very quickly. First of all, you let the Salvation Army know what a soldier needs and they take care of you. They took care of us in grand style. In addition, soldiers solved it as soldiers do everywhere, whether it was asking a pretty girl down the street for some help, or calling home and saying, deliver me such and so. They got their showers. I'm not aware of any hygiene problems that lasted for more than one or two days. That's one of the risks, by they way, when you scatter soldiers two-by-two. You've just got to count on them doing the right thing. The chaplains can't see them all and the docs can't see them all. Bless their hearts, as is so often the case, if you have good soldiers, they'll do what's right. And they did.

MOUT DOCTRINE PANEL PRESENTATIONS

ANNEX 1: USMC MOUT DOCTRINE (MCWP 3-35.3)
Maj Mark Sumner, USMC

USMC MOUT DOCTRINE

MAJ MARK SUMNER
DOCTRINE DIVISION, MCCDC

The topics that I will cover in this presentation include key elements of modern urban battle, the development highlights of the Marine doctrinal publication, the purpose, scope, and target audience, and key subjects covered in the publication.

MODERN URBAN BATTLE

- **Intelligence is imperative to success in urban warfare**
- **Surprise is a combat multiplier that can be achieved through deception, stealth, and ambiguity**
- **The MAGTF must continue to exploit the use of combined arms**
- **ROE and the "CNN factor" will continue to have impact on future urban operations**

The number and increasing size of cities throughout the world make it all the more probable that forward-deployed marines will fight in cities. Whether it is at the lower end of the range of military operations such as NEOs or high-intensity combat, marines must be prepared to fight effectively. Intelligence remains as the all-important key to operations, but perhaps more so in urban warfare. The complex environment, systems within systems, density of noncombatants, and cultural differences all drive the need for rapid and accurate intelligence. By using intelligence to drive operations, we can achieve tactical surprise while exploiting the combined arms of the MAGTF. Our marines and their commanders must also be cognizant of the influences of ROE and the CNN factor.

USMC DOCTRINE DEVELOPMENT

- **OPERATIONAL HANDBOOK (OH) 8-7, *MOUT*, DATED NOVEMBER, 1980**
- **RESEARCH FOR NEW PUBLICATION STARTED MARCH 1993**
- **INITIAL DRAFT COMPLETED DECEMBER, 1993**
- **COORDINATING DRAFT COMPLETED SEPTEMBER 1997**

We were using an older operational handbook for quite a while, but it provided us the detailed TTP that small-unit leaders were interested in. In 1993 we started the development of a new publication that would include a MAGTF commander's perspective on fighting the MAGTF on urbanized terrain. This new publication would include considerations for all four elements of a MAGTF conducting MOUT.

USMC DOCTRINE DEVELOPMENT

- **REVISIONS AND ADDITIONAL APPENDIX ADDED. MCWP 3-35.3 SIGNED APRIL 1998. SUPERSEDED OH 8-7.**

MCWP 3-35.3 was finally signed in April 1998.

MOUT DEFINED

- *All military actions that are planned and conducted on a topographical complex and its adjacent terrain where manmade construction is the dominant feature. It includes combat in cities, which is that portion of MOUT involving house-to-house and street-by-street fighting in towns and cities. [Marine Corps Reference Publication (MCRP) 5-12A, Operational Terms and Graphics]*

I wanted to bring out the point here that there is no approved joint definition for MOUT. This is the approved term used by both the Army and the Marines. When we discuss MCWP 3-35.3, this is the definition the Marines use. In my later presentation of JP 3-06, I will discuss a draft joint definition.

PURPOSE and SCOPE

- **Purpose: Provides doctrinal guidance and detailed information on tactics, techniques, and procedures to be employed by the operating forces in MOUT.**

- **Scope: It is written from a Marine air-ground task force perspective, with emphasis on the ground combat element as the most likely supported element in the urbanized environment.**

This is the purpose and scope. Emphasis is on the ground combat element.

TARGET AUDIENCE

- **Target Audience: Prepared primarily for commanders, staffs, and subordinate leaders down to the squad and fire team level.**

- *MCWP 3-35.3 primarily addresses MOUT at the tactical level of war.*

The target audience is the commanders and staffs, but we included a lot of detailed TTP so it would be useful to our small-unit leaders. MCWP 3-35.3 is primarily focused at the tactical level of war.

TOPICS COVERED

- **OFFENSIVE AND DEFENSIVE OPERATIONS**
- **COMBAT SUPPORT**
- **COMBAT SERVICE SUPPORT**
- **NONCOMBATANT CONSIDERATIONS**
- **MOOTW**
- **MOVEMENT, CLEARING, AND FIRING TECHNIQUES**
- **WEAPONS EFFECTS**
- **SUBTERRANEAN OPERATIONS**
- **LIMITED VISIBILITY OPERATIONS**
- **LESSONS LEARNED**

This is a list of the key topics covered.

KEY SUBJECTS

- Detailed discussion of weapons effects on urban structures.

- Recognizes the increase of urban areas, their center of gravity, and asymmetric threats.

- Incorporates lessons learned from recent events such as Somalia and Chechnya.

- Recognizes the implications of increased numbers of noncombatants; more restricted ROE; and the CNN factor.

Key subjects include a detailed discussion on weapons effects, recent lessons learned, and the implications and restraints placed on marines, including ROE and the CNN factor.

ANNEX 2:   USA MOUT DOCTRINE (FM 90-10)
LTC Mark Reardon, USA

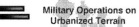 Military Operations on
Urbanized Terrain

AGENDA

- **Evolving Doctrinal Philosophy behind FM 90-10**

- **FM 90-10: New Perspective**

- **Going Beyond a Limited Tactical Approach**

- **Civilian Considerations**

- **Our Toughest Challenge - Offensive Operations**

- **Conclusion/Summation**

Meeting the Challenge

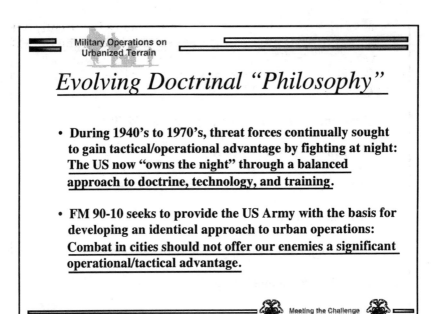

The U.S. Army has a proud tradition of overcoming, adapting, and improvising in response to evolving threats and battlefield dynamics. In the Second World War, both the Japanese and Germans developed sophisticated night offensive tactics to negate the impact of American airpower and supporting fires. This trend continued through the Korean War and Vietnam War.

During the 1960s, however, technology had advanced to a point where it could significantly enhance training designed to improve American combat performance at night. This process continued to evolve to a point where we arguably can state that "U.S. forces own the night," the result of a decades-long doctrinal, technological, and training solution to an asymmetrical approach employed by our adversaries.

I would submit that the U.S. Army, rather than continuing to espouse avoiding cities, should instead confront the question of urban combat as avidly as our predecessors did when they sought a solution to the question of night operations. The advantage that the Department of Defense has today is that the United States is currently not embroiled in a desperate urban struggle that is costing the lives of our servicemen on a daily basis. We can begin addressing this question *before* our adversaries can utilize urban warfare to gain a real advantage.

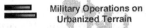

FM90-10: A New Perspective

- Current US Army doctrine developed in 1979 when we trained to fight a conventional conflict in Europe on allied territory.

- Current urban threats are not limited to a conventional type force.

- Our adversaries perceive they gain an advantage from fighting in cities.

- Global urban growth, when combined with US policy of engagement, increases potential for MOUT.

- "Rubble and clear" approach is no longer militarily or politically feasible.

- MOUT is much more than a question of firepower and technology - it must also encompass stability and support actions.

The current version of FM 90-10 was written when Active Defense and the General Defense Plan ruled supreme. The briefest of glances at the manual will disclose illustrations that feature typical German villages—a scene only slightly removed from our World War II experiences. Indeed, the Army's experience in that war, with the exception of Seoul and Hue (both USMC operations), formed the predominant foundation for the 1979 version of the MOUT manual.

Discussion of corps- and division-level offensive operations in the 1979 manual is concerned with bypassing or avoiding urban areas whenever possible. Brigade is the first echelon of command where specific discussion is found concerning combat operations within the town, yet even this discussion is focused on "systematic clearance of the town"—the World War II methods of conducting MOUT. During the discussion of defensive operations, the manual focuses only on the concept that MOUT is useful because it offers the defender a significant tactical advantage. Built-up areas are viewed as "obstacles" that should be integrated into the overall defensive scheme.

Things have changed immensely since 1979: the operational environment that we can expect to fight in no longer resembles that currently found in FM 90-10. No longer can we expect to combat a conventionally armed and organized adversary. It is immeasurably more difficult to "avoid" or "bypass" urban areas, and in most developed countries this has evolved to somewhat of an impossibility. Taking a cue from our doctrine, potential adversaries realize that they can accrue significant advantage from urban combat. This makes it very difficult for U.S. commanders to gain a force-oriented versus a geographical objective when the enemy deliberately chooses to fight in a city. The ability of the media to provide a center stage view of any military operation, when added to the popular and political expectation of "bloodless war," ensures that we can no longer conduct "rubble and clear" MOUT such as the Russians were forced to do in Grozny. The inevitable cost measured in time, casualties, and treasure demand that we seek a innovative solution to the question of combat in cities.

The current FM 90-10, which is largely focused on our World War II/Cold War paradigm, does not discuss stability actions and support actions—operations other than war requiring the employment of military forces that have been occurring with increasing frequency over the past several decades. Civilian considerations rate a page of discussion. In the current geopolitical environment, MOUT means much more than conducting combat operations, in an urban setting, against a conventional or unconventional foe. The responsibilities that come with maintaining *an entire city* are awesome, to the point that only the massive infusion of military resources can provide immediate aid to an urban area infrastructure that has collapsed as a result of natural catastrophe or civil disturbance.

Military Operations on
Urbanized Terrain

GOING BEYOND A LIMTED TACTICAL APPROACH

- MOUT expertise within the Army largely resides at the tactical level (platoon through battalion). This is largely due to the fact that MOUT facilities are not large enough to support brigade or higher operations nor do current simulations support training for MOUT at brigade through corps.

- Our tactical focus has obscured an important fact: MOUT is much more than conducting offensive and defensive operations; it will also include stability actions and support actions.

- The current Army focus on platoon through battalion level does not make full use of the capabilities of other services during MOUT.

- Limiting our doctrinal approach to the tactical level will not support our requirement to fight in "cities," nor does it dissuade those who still view MOUT as "business as usual" in an urban setting.

  Meeting the Challenge

The current FM 90-10 is aimed at the brigade through battalion level. This narrow focus has resulted in the conclusion of the 1994 Defense Science Board, following its review of MOUT, that "operational-level doctrine for MOUT is lacking." While a tactical focus may have been supportable in an era prior to "urban sprawl," it will no longer support nor inform division and corps commanders who may be required to focus all their resources on MOUT. The narrow tactical focus of FM 90-10 ensures that our doctrine does not include discussion of stability and support actions in a MOUT environment, since these are not missions that tactical units are tasked to undertake. Computer exercises, such as those conducted by the Army's Battle Command Training Program, rarely feature MOUT as part of an exercise. It is too tough a problem to replicate with our current generation of simulation devices. Even our MOUT training facilities are normally fairly small and unable to support larger units. All of these factors have resulted in MOUT expertise residing exclusively at the tactical level of our leadership—leaders at battalion level and below. Senior leaders have to rely on their experience at that level to aid them in planning and executing MOUT. Do we really want them to operate under that handicap? The draft FM 90-10 aims at filling the void that exists in terms of MOUT doctrine at the operational level by focusing on division and corps operations in a MOUT environment.

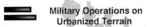 Military Operations on Urbanized Terrain

CONSIDERING THE CIVILIAN POPULATION

- Current Army doctrine characterizes the civilian population as "hapless spectators" - FM 90-10 must recognize that people are the lifeblood of a city and that they can have a decisive impact on MOUT.

- Turn asymmetrical "advantage" of MOUT against the threat. When he deliberately seeks to fight in cities, we should use IO to separate him from support that might normally be offered by the population.

- Civilian considerations form the basic foundation for MOUT planning. The use of fires, choice of maneuver "corridors," designation of nodal objectives, etc., are driven by civilian considerations. How can we effectively employ IO to separate threat from population if we inflict significant collateral damage? How can we support the needs of a 250K population when we have destroyed the city infrastructure?

 Meeting the Challenge

The first step toward revising our existing MOUT doctrine is to convince commanders that the population of a city, which may be larger than the respective military forces involved, will have an operationally significant impact on MOUT, far more than articulated in the current FM 90-10. The 1979 version devotes but a single page to the discussion of civilians in MOUT, buried within the CSS section of the manual. Civilians are seemingly characterized as an unavoidable nuisance; often they are seen merely as a means for the enemy to infiltrate disguised combatants through friendly lines. With this mind set, it is no wonder that our potential adversaries view combat in cities as working to their advantage.

The draft FM 90-10 will integrate emerging information operations doctrine with MOUT. Given the proliferation of media sources within most urban areas, the use of offensive IO, specifically tailored to address regional cultural and political concerns, can have a significant impact on the attitude of the civilian population. Has anyone ever tried to exploit the fact that civilians might not support a decision by their military to wage war against U.S. forces within their own cities? With the proliferation of media sources, cell phones, and modern communications in general, could we not realistically expect some civilians to aid our attempts to limit collateral destruction while quickly gaining our critical military objectives? Even within cities populated by hostile civilians, can we realistically expect them to have homogenous political and cultural attitudes? Offensive IO, conducted in a preventative and proactive manner, should convert hostile populations into neutral audiences as well as ensure that neutral populations do not become actively hostile.

With the trends in urban growth, we can no longer assume that cities can be avoided. We must remember that in many cases they will be larger and have commensurably increased populations. Support requirements for populations of this size will quickly have operational impact on friendly military resources. In response to this trend, commanders must take into account the desired end state in relation to the civilian population. In stability and support actions, this type of approach to MOUT goes without saying, for the population is often the primary focus of that type of urban operation. During combat, however, the population is normally not the immediate focus of the military commander. Yet we cannot ignore their needs, because civilians will eventually become the focus of the urban operation. If U.S. forces destroy power facilities and water plants during the initial phase of an operation, this means U.S. forces will assume responsibility for repair/maintenance of those services once the urban operation is concluded. For a 250,000-plus population, this requirement can consume considerable military resources. While support requirements should not constrain a commander, they should be considered and resourced before the urban operation begins.

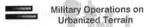

Military Operations on
Urbanized Terrain

Our Toughest Problem: MOUT Offensive Operations

A Possible Solution - FORCE SEQUENCING:

- To minimize collateral damage, mass is achieved by echeloning forces attacking along a narrow axis.

- Limited objectives are assigned to subordinate echelons within the attacking force. Follow-on echelons rapidly pass forward to maintain the speed and momentum of the attack until a complete penetration is achieved. The attacking force moves along the penetration "lane" to seize key nodal points within the city.

- Operations occur at a rate that is faster than the enemy can react in order to overwhelm, turn, and defeat him.

- Offensive IO seeks to ensure that the population remains neutral.

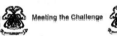

Meeting the Challenge

While the U.S. Army has not undergone a searing MOUT experience on the scale of Stalingrad, it can certainly borrow from the hard-won and expensive lessons of that campaign. A December 1943 article in *Infantry Journal* by Lieutenant General Chuikov of the Soviet 62nd Army provides the intellectual basis for an approach to operational-level MOUT: penetration on a narrow front at a rapid tempo to overcome the defenders' natural advantage. Mass is achieved by focusing effects on a narrow frontage. Momentum is sustained by cycling fresh forces continuously into the fight. Once a force passes through the lead element, it takes over the attack. The force that has just seized a city block immediately transitions to the defense to secure the penetration from lateral counterattacks. The "cycling of forces" process described by General Chuikov is extremely critical when one considers the physiological impact of MOUT on soldiers.

Advantages versus risk? Detailed reconnaissance by conventional forces and special operations forces must identify the time and location for the penetration attack. Without detailed reconnaissance and urban-specific IPB, these tactics will place friendly forces at risk.

Detailed reconnaissance and accurate IPB will facilitate precision maneuver, precision fires, task organization, offensive IO, and predicting the enemy's reaction to friendly operations.

Narrow frontages are susceptible to counterattack from the flanks as well as massed indirect fires. When the operational commander focuses his ISR assets to the flanks and designates the penetration corridor as a "critical friendly zone" for counterbattery assets, he is focusing their capabilities against an enemy forced to take active measures to combat his forces. No longer does the threat enjoy the cover and concealment traditionally provided by urban sprawl. Now he is forced to take to the streets and to utilize weapons systems that he might have considered crucial for a follow-on phase in the battle. Friendly forces, by echeloning assets to continually maintain their momentum, will begin to operate within the enemy's decision-making cycle. Adversaries will make decisions based on information that has already become obsolete—a condition that should be enhanced by offensive IO designed to attack enemy C2 nodes. The time-honored tactic of defending a city using concentric rings leading to a central position supports use of rapid penetration on a narrow front. The attacker could conceivably penetrate the inner rings before the defender has had time to reposition forces.

Penetration on a narrow frontage also addresses the matter of collateral damage. Let the enemy initially keep part of the city. Friendly forces will seize only the critical portions—using IO to their advantage to win the support of the population or suppress the activities of hostile elements.

The morale factors of isolation and fear should not be discounted. With friendly forces having penetrated the outer ring of defenses of a city, will those defenders in other portions of the urban area remain in their positions? Can we expect them to fight on even when surrounded and isolated from supplies? When bereft of guidance from higher headquarters? Indeed, some troops will possess the level of dedication and training to enable them to do this, but should we discount this factor in urban areas—terrain that naturally tends to separate units into small groups? Once again, when they begin retreating, of their own accord or as a result of orders from their higher command, friendly ISR assets target them for engagement by precision weapons once they are in the streets.

Combat forces, however, will have to be task organized in nontraditional ways to accomplish this so as to ensure that all the required tools are pushed down to the tactical level to lend the lead echelons responsive and effective support. The impact of this will transmit itself very quickly from the tactical level to the operational level. A light infantry company commander could find himself with a tank section, Bradley section, 155mm SP section, chemical smoke section, engineer platoon, ambulance section, CA team, MI interrogators, linguists, transportation section, and other assets. Task organizing multiple companies within a brigade or division in this fashion will rapidly overwhelm higher headquarters' organic resources. This will impact on how the operational commander views the flow of forces into theater. What is ostensibly a tactical problem becomes a deliberate decision at the operational level that impacts on the overall time phased force deployment data (TPFDD) sequencing.

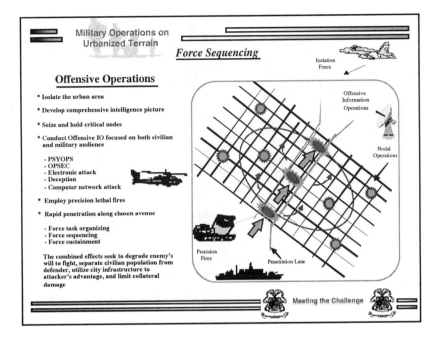

Many of the points discussed under the previous slide are documented in this example.

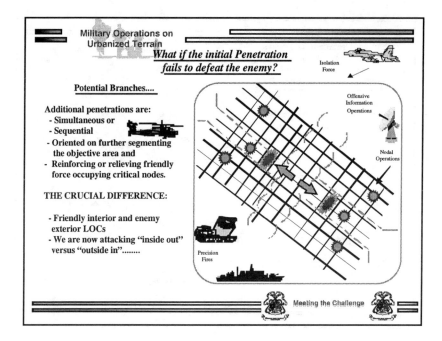

If the initial penetration does not result in complete success, one or more additional actions may be necessary.

Military Operations on
Urbanized Terrain

CONCLUSION/SUMMATION

- The US Army seeks to develop operational level MOUT doctrine that will drive detailed examination of related material, leader development, training, and organizational initiatives

- Develop offensive MOUT doctrine concept that will allow us to progress beyond the "rubble and clear" techniques of the past

- New FM 90-10 will recognize the unique IPB, cultural, and civilian considerations that will impact on MOUT

- Provides doctrine for stability and support actions in MOUT

- Seeks to broaden Army MOUT doctrinal perspective in relation to other services (train like you fight...)

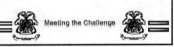
Meeting the Challenge

ANNEX 3: THE JOINT MOUT HANDBOOK
Maj Jeff Newel, USAF

JTF Commander's Handbook for Joint Urban Operations

Major Jeff Newell, J-8 Joint Urban Working Group

I'm really glad to be here today to bring you up to date on the J-8 Urban Working Group's handbook. When I took on this project for the joint Urban Working Group, I had no idea how many eyebrows would be raised at an airman's leading of the effort. Most people, of course, conclude that urban operations are primarily a concern of soldiers and marines. Last fall I truly discovered how rough the sailing would be when LtCol Duane Schattle, the UWG Chairman, appeared before the JRB, a senior policy board comprising general officers from each service. When he announced that the UWG had placed an airman in charge of this project, the Air Force (!) representative asked "Why in the heck is the *Air Force* writing a handbook on joint urban operations?"

As a member of the Joint Urban Working Group for almost a year now, and as the leader on the handbook project for almost that long,

I'm happy to give you a quick update on where we stand, and some idea about where I believe the handbook should fit in relation to the forthcoming joint urban doctrine.

Before I talk too much about the handbook itself, I think it's helpful to back up to the early fall of 1998 and briefly summarize the origins of the project and why I believe we are so far down the road on this handbook with so much support from the four services. Last summer, the UWG collectively decided that although we wanted a stand-alone doctrine publication on joint urban operations, this new doctrine might get bogged down in the writing process. As you know, contentious doctrinal issues sometimes reach the most senior levels before disputes are resolved, and many of us thought that urban doctrine might be one of those issues. In fact, I should be up front; the Air Force opposed the writing of a stand-alone, joint urban doctrine, and only time will tell whether Joint Pub 3-06 will sail smoothly through the ambitious, but achievable, two-year timeline. I hope so. But our UWG advisors from J-7 warned us that it might take twice that long before joint urban doctrine was ready for signature.

While few in the UWG really believed that doctrine would take that long, we generally agreed that with the collective knowledge gained by the group, we could produce a handbook that CINCs would find useful. The handbook would not take the place of doctrine, but it might offer CINCs a useful planning tool for the near term, something that could prep them for an urban operation, short of the authoritative guidance that the eventual doctrine will provide. So the UWG members decided that we could sponsor such a project and asked for a service to volunteer to lead, meaning "fund," the effort. Although the Air Force stepped forward, the handbook is really a UWG-sponsored project with the day-to-day research and writing in the able hands of Mr. Barak Cohen, a former Army officer, and Mr. Paul Nagy, a Naval Reserve officer. Both are representatives of DFI International and both are long-time contributors to the UWG. The UWG has collectively made all the key decisions on the project. These include designing the initial outline, choosing the case studies, recommending people to interview, and chopping on draft chapters as they were written. To date, we've received constructive inputs from all of the services. We look forward to the complete handbook draft being circulated for review in June 1999.

Now it's important to understand what the handbook is, and what it isn't. I get uncomfortable when someone occasionally says that the handbook is a "gap-filler" until the doctrine is complete, or in any way links it to the forthcoming joint urban doctrine. The primary reason for this discomfort is that doctrine writing is not in the purview of either the UWG or my Air Force organization, the Strategic Plans Directorate. In fact, we had to work very hard to assure the Air Force Doctrine Center, whose support I needed within the Air Force—that we weren't writing doctrine.

Well, if the handbook isn't addressing doctrinal issues, then what is it? To make this distinction between doctrine and what we were trying to do, we focused on the word *prescriptive* versus *descriptive*. Where doctrine is primarily *prescriptive*, telling the CINC how he *should* conduct an urban operation, we would attempt to write a *descriptive* guide to the urban environment, offering joint commanders and their staffs some idea of what to expect, and what might be different, when embarking on an urban operation.

Now I won't try to tell you that this division is a simple one, or that some descriptions don't have prescriptive implications. Let me offer an example. If I were to describe a city as a fragile organism relying on a few critical nodes to sustain its functioning, that *description* may drive one's thinking on how to conduct a military operation within it. By contrast, if I were to describe the city as a highly resilient and mutually supportive network of dispersed functions, that may drive you to a very different set of conclusions about how to conduct an urban operation, particularly about the wisdom of strategic air attacks against a city. The point here is that when writing the handbook, we've had to be mindful that we're not writing doctrine, yet what is or is not a doctrinal statement is not always clear. Members of the UWG have "kept us honest" in that regard, reminding us of our charter when it appears that we're getting too close to making a doctrinal statement or appearing to synchronize our handbook with the JP 3-06 draft. LTC Mike Beans of the Army Staff has been especially good about that, and I think the handbook will be better for it.

Before I get into the handbook itself, let me show you the timeline for the project so you can see exactly what stage we're at.

Handbook 1999 Timeline

√ January: Submit Handbook outline to the UWG
√ February: UWG reviews draft of Chapters 1 and 2
√ March: UWG reviews research on Big 5 case studies
• April: UWG review research on remaining potential cases
• Early May: submit draft of entire Handbook to UWG
• Late May: UWG submits suggested changes to draft
• Early June: UWG sends revised draft of Handbook to services and CINCs for 4 to 6 weeks of review
• Mid-July: UWG revises draft to incorporate CINCs' views
• Late July - August: final draft considered by UWG
• September: draft Handbook submitted for "136 Process"
• Mid-October?: final draft prepared for publication

As you look at this timeline, I think you'll be struck by just how fast this project is moving. The UWG approved the handbook outline in January; we hope to have the entire first draft back to the UWG for review in early May. Once the UWG has approved the draft, we will distribute it to the services and CINCs for comment. We'll use June through August for this vetting process, though frankly it's hard to say at this point how long this will take. Finally, we'd like to formally staff the handbook though the "136 process" for formal service approval, and have a final draft ready for signature in October 1999, one year from beginning the project. We hope to have the Director of the Joint Staff sign the handbook, and add the publication to the Joint library—similar to the *Joint Handbook for Peacekeeping Operations*. While it is likely that the JP 3-06, *Doctrine for Joint Urban Operations*, will eventually simply replace the handbook in the joint library, we'll have to wait and see whether the joint doctrine obviates the benefits of the handbook. At this point, I think it's an open question.

Now let me walk you through the outline for the publication.

Handbook Outline

1. Introduction to Joint Urban Operations

2. The Operational Context of Joint Urban Operations

3. Practical Considerations in Joint Urban Operations

4. Case Studies

As you can see, the handbook consists of four chapters, with the final chapter consisting of five case studies. Keep in mind that the focus of this publication is at the operational level of war. We've taken pains to stay away from tactical issues.

Chapter One is an overview of the publication, defining its scope and purpose: "pending the publication of joint doctrine for urban operations, this handbook provides JFCs, their staffs, and other interested parties within the Department of Defense with a primer on joint urban operations." Supplementing existing doctrine that guides the conduct of military operations in urban terrain—namely JP 3-0, *Doctrine for Joint Operations*, and JP 3-07, *Joint Doctrine for MOOTW*—this handbook addresses explicitly "the issues and considerations that JFCs may wish to review during the planning and conduct of joint urban operations."

Chapter Two attempts to place urban operations within the appropriate operational context. It makes the distinction between joint urban operations and its subset, urban combat. This chapter also addresses how an urban operation should fit within larger, operational-level plans. It simply makes the point that in some operations, striking directly at the urban area is critical to achieving the operational objectives of the campaign. Panama is a good example. Other times, the urban area may not be central to achieving the larger operational objective, and might be bypassed. In addition, Chapter Two makes the case that it is at the operational level that the decision is made to engage in urban combat. Said another way, it is operational-level commanders who place forces in a position that may require them to engage in urban combat.

I believe Chapter Three, "Practical Considerations in Joint Urban Operations," will be the most substantive and useful. This chapter attempts to make some distinctions between differing types of urban environments and addresses some intelligence considerations in urban areas. This chapter also discusses civil-military considerations, multinational coalitions, ROE, and interagency coordination. A substantial amount of research, including personal interviews, has gone into getting this chapter right.

Finally, Chapter Four consists of five case studies, which are essentially two- or three-page summaries of the particular operations, and the subsequent lessons learned.

Proposed Case Studies and Vignettes

- Algiers (1957)
- Baghdad (1991)
- Beirut (Israel, 1982; US, 1983)
- Budapest (1945)
- Freetown (1997)
- Hue (1968)
- Jerusalem (1967)
- Khafji (1991)
- Tirana (1997)
- Monrovia (1996)
- Manila (1945)

- Belfast (1980s - present)
- ➡ **Grozny (1995)**
- Berlin (1947-48)
- Kuwait City (1991)
- ➡ **Mogadishu (1991, 92-93)**
- ➡ **Panama City (1989)**
- ➡ **Port au Prince (1994)**
- ➡ **Sarajevo (1992-95)**
- Stalingrad (1942)

As you can see by the list, we did some preliminary research on several case studies, looking for lessons relevant to today's joint commanders. After discussing the issue within the UWG, we decided that we had to say something about what we call "The Big Five," indicated with arrows. All are post–Cold War examples in which the urban operation was central to achieving operational objectives. We are continuing to look at the other urban operations listed to see if they are worthy of inclusion in this chapter. We will probably include some of these as vignettes placed throughout the publication.

Finally, let me quickly list our group of reviewers.

Reviewers

- **Ambassador Robert B. Oakley, US Foreign Service, Ret**
  - Special Envoy for Somalia, Dec '92-Mar '93, Oct '93-Mar '94
- **Admiral Gary Wheatley, USN, Ret**
  - Former CO USS John F. Kennedy, former director of operations for command and control for the Atlantic Fleet, former Senior Fellow at the Hudson Institute, co-edited *Interagency and Political-Military Dimensions of Peace Operations: Haiti--A Case Study*
- **Brigadier General Richard Potter, USA, Ret**
  - Former CO, 10th Special Forces Group (Airborne), and Deputy Commander, US Army Special Operations Command
- **Colonel Tim Thomas, USA, Ret**
  - Foreign Military Studies Office, Fort Leavenworth, Kansas
- **Colonel RW Trost, USA**
  - Chief, Information Strategy Division, J3 (J39-ISD)
- **Dr. Kathy Fine-Dare, Professor of Anthropology, Fort Lewis College**
- **John Cope, Senior Military Fellow, NDU**

These are various military and civilian experts who have either some particular expertise relevant to the project or experience as a joint force commander or planner. I should update that list to include Major General Ernst, Commandant of the U.S. Army Infantry School at Fort Benning, Georgia, who has taken an active interest in the project. He graciously gave us a couple of hours last week to comment on our draft and suggest improvements.

Reviewing Agencies

- Department of Justice
  - Daniel S. Seikaly, Director, Executive Office for National Security
- Department of the Treasury
  - Peter J. Dowling, Special Agent in Charge, Liaison Division, Government Liaison and Public Affairs, US Secret Service
- Department of Transportation
  - Commander Peter J. Ganser, USCG, Manager, Policy and Operations, USCG HQ
- Federal Emergency Management Agency (FEMA)
  - Michael Austin, National Security Coordination Office, FEMA
- Peace Corps
  - Gloria Johnson, Director, External Relations Office, the Peace Corps
- US Agency for International Development (USAID)
  - Julie Klement, Foreign Disaster Assistance Office, USAID

In addition to these reviewers, we've got various government agencies looking at the drafts to give us an outside perspective. We all agree that future urban operations will typically involve more than simply the DoD, so we think it's important to get the interagency perspective right from the beginning.

JTF Commander's Handbook for Joint Urban Operations

Major Jeff Newell, J-8 Joint Urban Working Group

I encourage the various service representatives to pass the handbook draft around when the Urban Working Group sends it out for review in early June. Your inputs are critical to ensuring that we get out the best possible product. It will also ensure that no one gets surprised when we gin up the formal staffing process in September. Thanks to those of you who have already made substantive comments to the handbook, and on behalf of the urban working group, I hope this publication contributes to building a more effective joint fighting force in urban terrain. I look forward to questions and comments during the panel discussion. Thanks.

ANNEX 4: USMC MOUT AVIATION DOCTRINE
Maj Floyd Usry, USMC

MARINE AVIATION WEAPONS AND TACTICS SQUADRON ONE

OVERVIEW

- •MAWTS-1 AVIATION URBAN INITIATIVES
  - •ACE MOUT MANUAL
  - •CAS SURVEY
  - •USMC URBAN CAS STUDY
  - •URBAN TRAINING CENTER
  - •URBAN CAS ASSESSMENT
- •QUESTIONS

MAWTS-1 has five initiatives in accordance with the MAWTS-1 Plan of Action Milestones (POAM) for aviation urban operations. This POAM is based on a lack of quantifiable testing of Marine close air support operations on urban terrain. It begins by looking at aviation urban techniques outlined in the MOUT manual, followed by a poll of CAS users and providers, followed by an outside, but formal view of CAS deficiencies and potential solutions. Finally, an Aviation Urban Training Center will be built and an Urban CAS Assessment conducted in Yuma, Arizona under the management of MAWTS-1.

MAWTS-1 MISSION STATEMENT

".....To provide standardized advanced tactical training and certification of unit instructor qualifications in all aspects of the tactical employment of Marine Aviation Units; and to provide assistance in the *development and employment of aviation weapons and tactics.*"

Emphasizing the development and employment of aviation weapons and tactics is our charter and impetus for conducting an Urban CAS Assessment (UCA).

MAWTS-1 ACE MOUT MANUAL

- **EIGHTH EDITION**
  - SECOND ELECTRONIC EDITION
- **INPUT FROM ...**
  - USMC, USA, USAF, WTI, SOTG, MCIA
  - WWII, SOMALIA, CHECHNYA, KHAFJI
- **SCOPE**
  - MAGTF STAFF TO "TRIGGER PULLER"

The MAWTS-1 Aviation Combat Element (ACE) MOUT Manual is updated twice a year. It is available on CD-ROM with embedded video. It is a compilation of "techniques" from urban conflicts around the world. Sources are numerous, to include most U.S. services and many foreign countries. It is written principally for the junior aviator who may be readying himself to launch into the next urban conflict, but it has applicability for staff officers as well.

MAWTS-1 ACE MOUT MANUAL

- **16 CHAPTERS AND APPENDICES**
  - 6 FUNCTIONS OF MARINE AVIAITION
    - EMPHASIS ON OAS (CAS)
  - MOOTW
  - C2W
  - TARGETING
  - IR POINTER EMPLOYMENT
  - ROE

The manual is composed of sixteen chapters and attempts to cover many of the factors inherent in urban operations, to include infantry, artillery, and aviation combat techniques as well as others such as the human dimension.

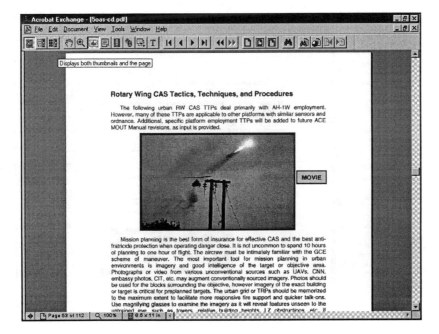

The manual has a strong emphasis on offensive air support and close air support from a rotary wing perspective.

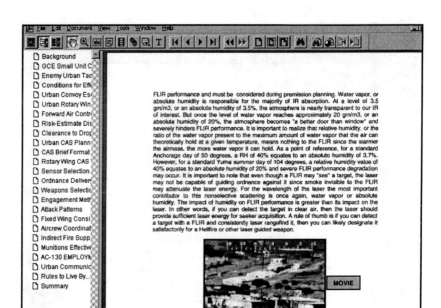

Close air support, as opposed to deep air support or air interdiction, is emphasized due to the complexity of supporting ground forces.

MAWTS-1 CAS SURVEY

SURVEY OBJECTIVES
- Ask users, "How are we doing with providing CAS in general?"
- Explore user opinions on UCAS issues
  - Targeting, C3, Fire Support & SEAD, Weapons and Munitions Effectiveness
- Validate research focus
- Compare ratings of Primary CAS Users/ Primary CAS Providers.

The CAS survey was meant to poll Marine combined arms officers about their attitude regarding Marine CAS, especially CAS performed in an urban environment. The survey results helped to validate the research focus for the formal Urban CAS Study. Basically, primary CAS users and CAS providers were the principal focus in the results. However, all comments in the survey were accepted and analyzed.

MAWTS-1 CAS SURVEY

- **MAWTS-1 CAS Survey launched July 97**
  - 64 Questions
  - 800+ distributed: MCU, Divisions, Wings, EWTGs, SOIs

- **Data cutoff - 31 Dec 97**
  - 421 Completed Surveys Returned

- **Data Analysis complete**
  - 26,880 numerical & 2,886 written data points transcribed and analyzed.

- **Final report completed**

  "Who participated?"

The CAS survey included 64 questions. Over 800 surveys were distributed to various Marine units. Four hundred and twenty-one surveys were returned, and over 2,800 written responses to questions were analyzed by MAWTS-1.

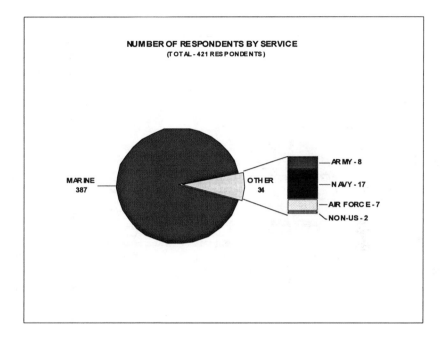

NUMBER OF RESPONDENTS BY SERVICE
(TOTAL - 421 RESPONDENTS)

MARINE
387

OTHER
34

ARMY - 8
NAVY - 17
AIR FORCE - 7
NON-US - 2

The principal respondents were Marine officers, who numbered 387.

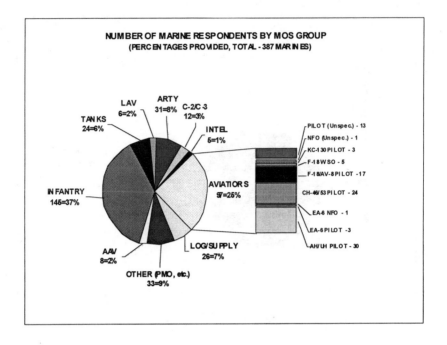

Infantry respondents accounted for 37 percent and aviators accounted for 25 percent of all responses.

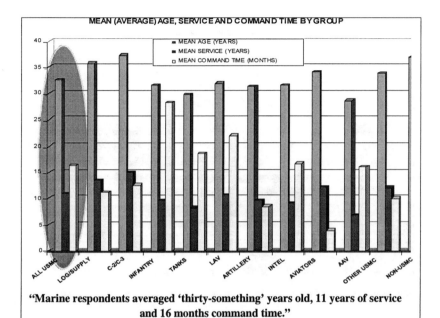

"Marine respondents averaged 'thirty-something' years old, 11 years of service and 16 months command time."

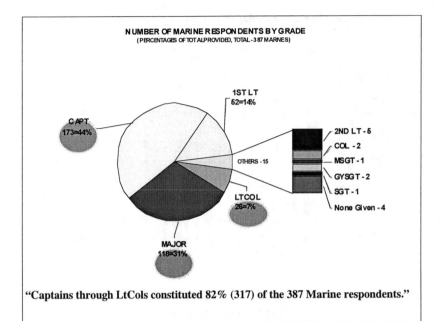

NUMBER OF MARINE RESPONDENTS BY GRADE
(PERCENTAGES OF TOTAL PROVIDED, TOTAL - 387 MARINES)

1ST LT
52=14%

CAPT
173=44%

OTHERS - 15

2ND LT - 5
COL - 2
MSGT - 1
GYSGT - 2
SGT - 1
None Given - 4

LTCOL
26=7%

MAJOR
118=31%

"Captains through LtCols constituted 82% (317) of the 387 Marine respondents."

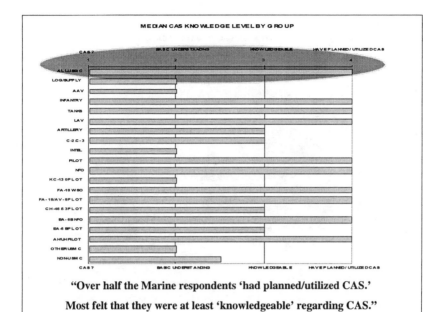

"Over half the Marine respondents 'had planned/utilized CAS.'

Most felt that they were at least 'knowledgeable' regarding CAS."

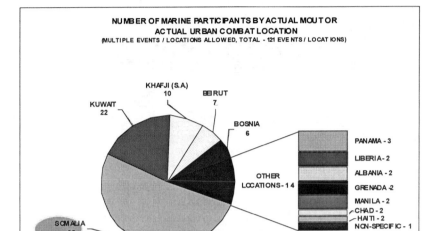

NUMBER OF MARINE PARTICIPANTS BY ACTUAL MOUT OR ACTUAL URBAN COMBAT LOCATION
(MULTIPLE EVENTS / LOCATIONS ALLOWED, TOTAL - 121 EVENTS / LOCATIONS)

KHAFJI (S.A.)
10

BEIRUT
7

KUWAIT
22

BOSNIA
6

OTHER
LOCATIONS - 14

SOMALIA
62

PANAMA - 3
LIBERIA - 2
ALBANIA - 2
GRENADA -2
MANILA - 2
CHAD - 2
HAITI - 2
NON-SPECIFIC - 1

Of the 121 Events/Locations cited by Marines who participated in MOUT, to include actual urban combat, most (62) occurred in Somalia.

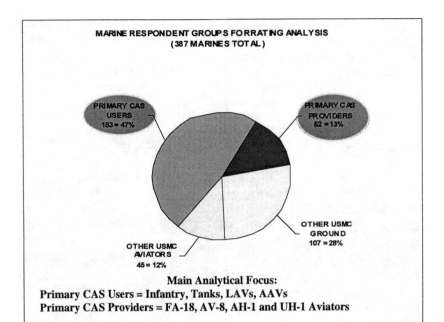

MARINE RESPONDENT GROUPS FOR RATING ANALYSIS
(387 MARINES TOTAL)

PRIMARY CAS
USERS
183 = 47%

PRIMARY CAS
PROVIDERS
52 = 13%

OTHER USMC
GROUND
107 = 28%

OTHER USMC
AVIATORS
45 = 12%

Main Analytical Focus:
Primary CAS Users = Infantry, Tanks, LAVs, AAVs
Primary CAS Providers = FA-18, AV-8, AH-1 and UH-1 Aviators

Conclusions (from responses)

- Overall, the respondents thought CAS provided by Marines was the best of all services.
- They also saw CAS Targeting, C-3, Fire Support and SEAD, and Weapons and Munitions Effectiveness in an urban environment as different than in other environments - requiring special consideration and preparation.
- In the ratings and comments *the respondents* found:
  - a) Marine CAS Doctrine to be *adequate* for all environments;
  - b) the organization for CAS control to be *inadequate* for the urban environment;
  - c) the CAS training range realism, training ordnance stocks and training frequency to be *inadequate* for all environments;
  - d) Current CAS ordnance yield to be *incompatible* with urban missions and
  - e) Facilities and support for training for CAS in an urban environment to be *inadequate*.

Recommendations (MAWTS-1)

- **Doctrine:** Use the annual JCAS Symposium process to clarify or reinforce Joint CAS Doctrine and definitions;
- **Organization:** Examine the USMC CAS controller organization carefully and determine what level is required for the urban environment;
- **Training:** Standardize *all* CAS controller training and review USMC training ordnance allowance levels for adequacy;
- **Equipment:** Develop/procure the scalable yield urban CAS weapon and other equipment required for the urban fight; and
- **Support and Facilities:** Another service or DoD build a second Aviation Urban Target facility in the eastern U.S. and upgrade all DoD operational simulators to Distributive Interactive Simulation (DIS) standards.
- **TTPs:** Upon completion of the Aviation Urban Target Complex at Yuma, MAWTS-1 will conduct a complete *Urban CAS Assessment* of current USMC CAS capabilities and TTPs.

Recommendations (MAWTS-1)

- The *CAS Quality Survey* report provides a baseline of Marine opinions concerning Marine CAS and, more specifically, Marine CAS in an urban environment. This report should be included in all future Marine and joint studies of CAS and MOUT.

USMC Urban CAS Study

- **Proposal Initiated By MAWTS-1 Sept. '97**

- **Study Approved: Marine Corps Study System Jan 98**
  - MAWTS-1 as Study Sponsor
  - Fleet Representatives In Study Advisory Committee (SAC)

- **First Study Advisory Committee (SAC) Meeting Jan 98**
  - Phases and Tasks reviewed and amended.

The survey highlighted the differences in operating on urban versus other terrain. It also emphasized that Marine aviators may support forces other than those in the USMC, such as our nation's soldiers.

Going outside Marine aviation required a formal study because it was beyond the MAWTS-1 framework for informal tactics development and evaluation. MAWTS-1 is the study sponsor and ORC is the study performer.

USMC Urban CAS Study

- **Objectives**
  - Comprehensive literature search
  - Identify USMC UCAS deficiencies, provide potential DOTES solutions
  - Develop methodologies for evaluating selected deficiencies and solutions
  - Develop complete POAM for follow-on Urban CAS research
    - Facilities construction requirements input
    - Modeling and Simulation requirements input
    - Test plan MOE input

- **Study Kickoff Meeting 28 August 98**
- **Study Final Report Completion May 99**

The study identified 43 deficiencies. MAWTS-1 has focused on the 17 top deficiencies requiring specific methodologies to address them. The desired end state is to have a complete report on the urban CAS problem that marines at all levels can reference now and in the future for improving our warfighting capabilities.

URBAN TARGET COMPLEX

- MAWTS-1 PROJECT MANAGER
- MWSS-371 CONSTRUCTION
- AUGMENTS BY I MEF
- MCAS YUMA
- NAVAIRPAC ENGINEERING ADVICE

The UTC's Project Office is MAWTS-1 ADT&E. MWSS 371 has built a representative slice of urban terrain unlike any other facility in the world. Welders were provided by I MEF. The project was supported and approved by MCAS, Yuma and NAVAIRPAC Tactical Target Division.

URBAN TARGET COMPLEX

- **FIRST IN DOD**
- **SCALEABLE DESIGN**
  - **167 BUILDINGS**
  - **1-5 STORIES (35')**
  - **CONTAINER BUILDINGS**
    - **CLUSTER BOMB UNIT SHIPPING CONTAINERS**
    - **OCEAN CARGO CONTAINERS**

The UTC is the first in DoD. It has 167 scorable buildings up to five stories or 35 feet. The buildings are made from cluster bomb unit shipping containers and ocean cargo containers arranged to represent real structures.

Cluster bomb unit containers stacked four high.

URBAN TARGET COMPLEX

- **FEATURES**..................
  - – VIRTUAL INTERACTIVE TARGETS
  - – NO DROP WEAPONS SCORING SYSTEM
  - – WEAPONS IMPACT SCORING SYSTEM
  - – GROUND AND AIR LASER CERTIFIED
  - – FORWARD AIR CONTROLLER TOWERS
  - – STREETS AND STREETS LIGHTS
  - – PAINTED BUILDINGS
  - – MILITARY AND COMMERCIAL VEHICLES
  - – ARTIFICIAL WINDOWS

VIT will provide damage assessments on buildings. Although container buildings, each will represent a real-world texture (brick, block, etc.) and function (school, apartment, etc.). VIT will provide damage assessments on the interior and exterior, in effect providing intentional and nonintentional damage. NDBS will score electronic ordnance drops via the TACTS. WISS will score actual ordnance drops. The entire complex is laser capable from 360 degrees. There are 4 FAC observation towers, 22 streets, and 12 working streetlights around the complex. Buildings are painted to represent brick or concrete. Thirty-four vehicles such as tanks, APCs, and cars are in the UTC.

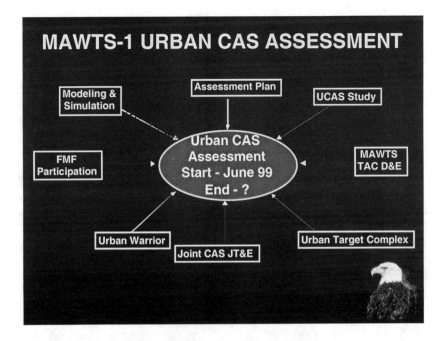

The Urban CAS Assessment will build on the Joint CAS T&E and the USAF 422 TES Urban CAS Study. It will assess the capabilities of the USMC's FA-18, AV-8, AH-1W and UH-1N aircraft, pilots, and TACPs to perform urban CAS using the Urban Target Complex as representative terrain. It will be flown by pilots from across the USMC. The UCA will build on the lessons learned from Urban Warrior and will support modeling and simulation via the model, test, model concept. It will start in June 1999 and run through December 1999 or until complete.

ANNEX 5: USAF MOUT TESTING
Maj Brooks Wright, USAF

TACTICAL AIRPOWER APPLICATIONS IN URBAN CLOSE AIR SUPPORT (UCAS)

THIS BRIEFING IS UNCLASSIFIED

That's me. "Kato," which is exactly like "Maverick" in *Top Gun*, is my call sign (only different).

We're going to talk about close air support.

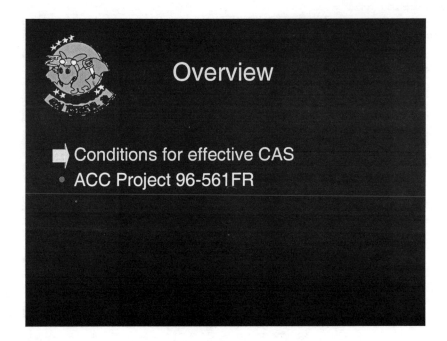

The presentation has two primary sections. We will first look at the conditions necessary for effective close air support.

Conditions for effective CAS

- JP 3-09.3 JTTP for CAS (pg I-5)
  - Air Superiority
  - Suppression of Enemy Air Defense
  - ➡ Target Marking
  - Favorable Weather
  - Prompt Response
  - ➡ Aircrews and Terminal Controller Skill
  - ➡ Appropriate Ordnance
  - Communications
  - Command and Control

These are all the things that go into effective close air support according to the joint manual. The three noted by arrows are the things that we attempted to address in a test I conducted.

A/OA-10

- **Primary Functions**
  - A-10: Close Air Support
  - O/A-10: Airborne Forward Air Control
- **CAS Capabilities**
  - Focused mission/highly trained pilots
  - Long loiter time
  - Strong, survivable airframe
  - LASER Spot Tracker to assist ID and LBG employment
  - Large weapons payload
- **CAS Limitations**
  - Not all weather
  - No targeting pod to designate for LGBs

That's the aircraft I fly. Those are the things it can do and some of the things it cannot do. This test was designed by my peers so that we could give them the information they needed. It does not include other aircraft.

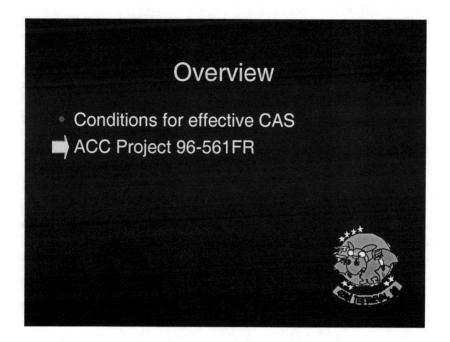

That's the test as it is titled. For those of you who seek it, you can go through ACC to get the test and they'll give it to you on CD-ROM.

Operational Necessity

- Operation JUST CAUSE in Panama
- Battle of Khafji during Operation DESERT STORM
- Task Force RANGER in Mogadishu, Somalia
- Operation JOINT ENDEAVOR in Bosnia

These were the reasons that the test came about. The last bullet is the result of my peers and A-10s holding over places like Sarajevo saying, "Here we are. We weren't trained for this. How are we supposed to skin this cat?"

Resulting in...

- ACC TIP 95-55: CAS and ASC in MOUT
  - ID and marking of targets and friendlies
  - Weapons and deliveries to limit collateral damage
- ACC Project 96-561FR
  - "Urban Close Air Support Tactics Development and Evaluation"
  - Final report dated April 1998

That resulted in a tactics improvement proposal that asked "How do we do this?" Two and a half years later, the answer came from the test that we generated and reported on. Those were the problems that we wanted to solve for operators; they addressed the questions they had asked us when they flew over urban environments.

That's the name of the test. That's the final report. It is classified SECRET.

Test Objectives

* Provide tactics for target talk-on, acquisition, identification and marking UCAS targets
* Provide tactics necessary to employ ordnance against UCAS targets
* Determine suitable munitions for minimizing collateral damage in urban terrain

That's the scope of the test. Those are the places we flew during it. There were limitations in some places because of the air space available to us for training or testing.

Test Parameters

- MOUT environments for CAS
  - FM 90-10-1 Infantryman's Guide... (May 1993)
    - Core periphery (European style with little high rise)
    - Residential sprawl (suburbia or Third World)
- Threat Assumptions
  - Localized air superiority
  - Preponderance of small arms/automatic weapons
  - Camouflaged optical light to medium AAA (23-57 mm)
  - Man portable IR SAMs (SA-7/14/16)
  - Few to no mobile IR or radar SAM systems (SA-6/8/13)

These were the things we wanted to do for our peers. Excuse the acronym "UCAS." I use that a lot. I'm not trying to create more acronyms, but urban close air support was the entire focus of this test.

Recommendations
"Mission Prep"

- Establish a "training program" for units subject to MOUT tasking
- TACs use mirrors and directional strobes to mark position
- Procure appropriate topographic maps, street maps and satellite imagery for potential MOUT areas

We have ordnance. How do we use it in an urban environment?

Recommendations
"Ordnance"

- Test and evaluate marking ordnance in MOUT environment
- Test and evaluate munitions effects in urban environment
  - Expand Appendix G (Risk-Estimate Distances) of JP 3-09.3 to include findings
- Provide AGM-65G and field AGM-65 H/K for UCAS

Which ordnance should we use and is the current inventory satisfactory?

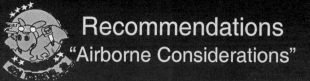

Recommendations
"Airborne Considerations"

- Conduct UCAS during daylight, threats permitting
- Use medium altitude tactics, degrade to high altitude and last, low altitude
- Assign FAC(A)s to support TACs during UCAS

We had six different urban environments available for testing. We chose the two we thought we were most likely to encounter and narrowed our scope to those two areas because we duplicate those on test ranges.

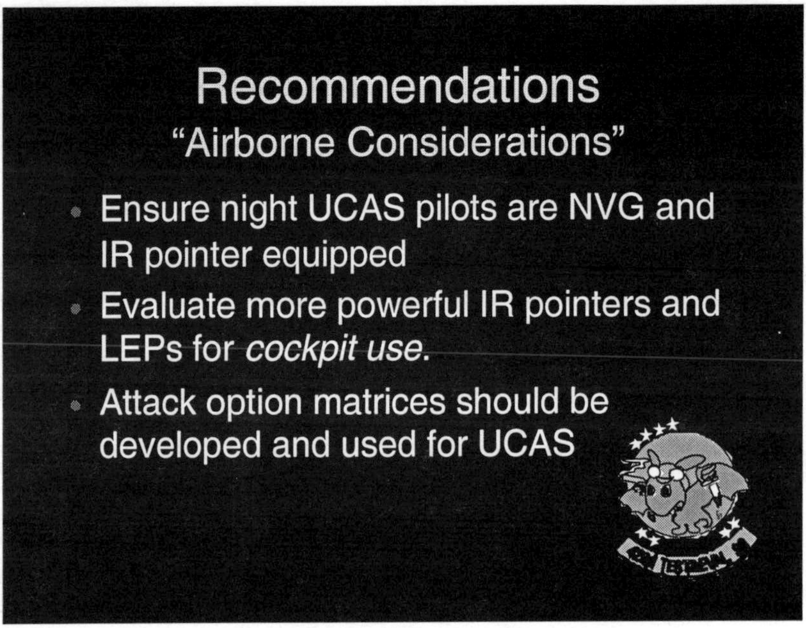

We had to make some assumptions in this test.

Recommendations
"Target Acquisitions"

- Pave Penny (LST) not used as *sole* means of target ID, but as a cueing device
- Use the same sensors for target identification and weapons employment
- Re-verify target identification prior to using stand-off weapons

These are the things that came out of this test. They're not in any specific order. They are grouped by subject matter.

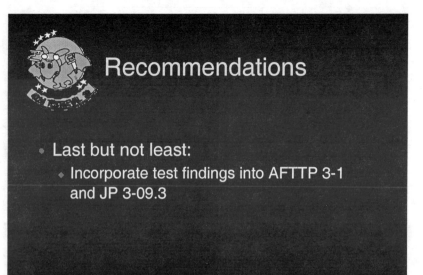

Conditions for effective CAS

- **JP 3-09.3 JTTP for CAS** (pg I-5)
  - Air Superiority
  - Suppression of Enemy Air Defense
  - Target Marking
  - Favorable Weather
  - Prompt Response
  - Aircrews and Terminal Controller Skill
  - Appropriate Ordnance
  - Communications
  - Command and Control

We have people on the ground. We need to be able to identify them before we employ ordnance. This is not new for those involved in close air support.

Air Force Tactics, Techniques, and Procedures 3-1 (AFTTP 3-1) is the bible for the mission design series (MDS) A-10, F-16, and others that the Air Force guys fly. If I'm going to teach Air Force guys how to do this mission, it needs to be in the air. If I'm going to talk to the rest of the communities about this mission, it needs to be in a joint pathway. Somebody needs to take charge of that. Hopefully Jeff [Newell] and his people [on the Air Force staff] will get involved in that.

Conclusions

- UCAS is difficult but possible
  - Need enhanced training opportunities
    - Get more MOUT sites
    - Standardize maps/imagery
    - Assess ordnance in MOUT environment
  - Update/incorporate lessons learned in current publications
    - Update JP 3-09.3 and AFTTP 3-1

This was a real weak area. We talk a lot about all of the capabilities we heard about—Predator UAVS taking photographs, overhead satellite imagery handed to the warfighter. Well, ask how many grunts on the ground have that map in their hands and you'll find it's very few. We found that if we're going to talk to those on the ground, whether a Delta Force operator, a Navy SEAL, a marine, or an Air Force guy sitting there with the Army, we all need to have the same maps. Currently, that is a definite hole in our planning process.

Those were the three things we looked at. We've made recommendations. The 500-pound bomb series and smaller is pretty good. Above that is probably too big. We weren't able to do target marking, so that was a deficiency.

We can do it. We may not be the best thing in the world right now. We may not be as good as the Marines at close air support, but we can get the job done if we have to.

We do need to train, though, and that is a real issue for those of us who are on the tip of that spear.

Maps and imagery are important. We need more ordnance.

Whatever we learn, let's put it in a format that everybody can access, not keep it in an Air Force stovepipe and not share it.

Contact Information

- **Maj Wright/ 422 TES**
  - DSN 682-7295
  - Comm (702) 652-7295
  - wright.brooks@nellis.af.mil

ANNEX 6: JOINT MOUT DOCTRINE (JP 3-06)
Maj Mark Sumner, USMC

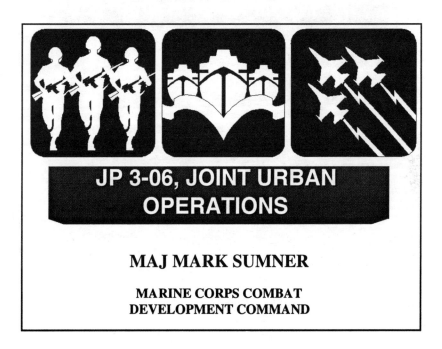

Good afternoon, ladies and gentlemen. My name is Major Sumner. As the action officer for JP 3-06, I will provide a presentation today on the publication's development.

AGENDA

- BACKGROUND
- CONCEPT
- JP 3-06
- SUMMARY

The agenda for this brief is as listed.

JP 3-06 BACKGROUND

- **RECOMMENDED BY J-8**

- **APPROVED BY JOINT DOCTRINE WORKING PARTY IN OCTOBER, 1998**

- **USMC ASSIGNED AS LEAD AGENT**

- **US ARMY ASSIGNED AS TECHNICAL REVIEW AUTHORITY**

- **J-8 ASSIGNED AS SPONSOR**

I think most of you are familiar with the J-8 Urban Working Group. The UWG, with the agreement of most of the services, recommended the development of a new joint publication focusing on the operational level of urban operations. This was agreed upon at a meeting of the Joint Doctrine Working Party in October 1998. The Marines were assigned as lead agent, with the Army as technical review authority.

Members of the UWG also decided to develop a new term called *Joint Urban Operations* with a definition of all services would agree on. This definition will first be staffed in the JP 3-0 revision.

Two significant differences between this definition and that for MOUT are:

- The joint definition identifies the density of noncombatants as a dominant feature of urban areas, and

- It excludes reference to house-to-house fighting.

JOINT URBAN OPERATIONS

- All joint operations planned and conducted across the range of military operations on a topographical complex and its adjacent natural terrain where man-made construction and the density of noncombatants are the dominant features. (draft definition)

CONCEPT

- **DEVELOPED UNDER GUIDANCE OF URBAN WORKING GROUP (J-8)**

- **USMC ASSIGNED TO INCORPORATE EXISTING ARMY, AIR FORCE, AND MARINE CONCEPTS**

- **EVALUATED DURING J-8 URBAN SEMINAR/WARGAME, SEPT 1998, AND USED BY SOCOM WARGAME, FEB 1999.**

Once the J-8 UWG decided upon the development of a new publication, we discussed the need for developing a concept or framework from which the publication could be developed. This was done by combining existing service concepts and testing the draft concept during war games.

OPERATIONAL CHALLENGE

- **Noncombatants & Infrastructure**

- **Complex, multifaceted operations**

- **Coordination and communication**

  - **US Govt, DOD, agencies**
  - **Allied governments**
  - **PVOs, NGOs**
  - **Multinational corporations**

The J-8 UWG felt that because of these challenges faced by a future JTF commander, a new framework or concept should be developed.

IMPACT OF NONCOMBATANTS

- "The number of Kosovars leaving, or being forced out at gunpoint, has escalated sharply since Wednesday and by midday Thursday had reached around 160,000. The largest number, more than 100,000 persons, continued to arrive in Albania, in particular the town of Kukes." UNHCR report, 1 Apr 99

Although not an urban operation per se, recent events in Kosovo highlight the difficulties that a JTF commander would face with non-combatants when conducting urban operations in such megacities as Seoul.

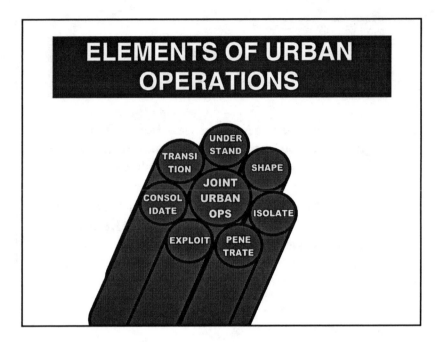

This concept is designed so that it is not conducted sequentially, but done simultaneously with specific points carrying the weight of the effort at a given time.

JP 3-06 SCOPE

- Provide fundamental principles and doctrine for the conduct of joint and multinational operations in urban areas.

- Focus on the operational level of war for the Joint Force Commander (JFC) and on the planning, training, and application of force across the range of military operations.

The scope and focus of the publication are listed here. Let me point out that it is not the intent of JP 3-06 to delve into the tactical level of war.

JP 3-06 TARGET AUDIENCE

- **COMMANDERS OF COMBATANT COMMANDS, SUBUNIFIED COMMANDS, JOINT TASK FORCES, AND SUBORDINATE COMPONENTS OF THESE COMMANDS.**

This target audience is in response to a survey conducted by the J-8 of former JTF commanders. It also included Ambassador Oakley, who said that this joint publication needs to be developed.

JP 3-06 KEY SUBJECTS

- FUNDAMENTALS
- OPERATIONAL TASKS
- NONCOMBATANTS
- INFRASTRUCTURE
- TRAINING
- URBAN IPB
- JOINT FIRES
- AIRSPACE CONTROL

These are the key subjects to be addressed in JP 3-06. As we continue to research and write we will probably find that we'll have to tie into other publications or perhaps propose new publications in order to keep JP 3-06 at a reasonable level of detail.

METHODOLOGY

- **Advisory contract with RAND to provide research.**
- **Attendance of Joint and Service conferences.**
- **Discussion with various CINCs.**
- **Europe trip planned in June, 1999.**
- **Progress tracked by J-8, UWG.**

We believe our methodology and research will be detailed. The Marines have hired a former Army officer who is a writer and Ph.D., Dr. Chris Bassford.

Additionally, we have contracted RAND, who will provide a team of experts to conduct research and review drafts of the publication. Progress is tracked monthly by a Joint Advisory Committee, who report back to the J-8 UWG. This, we hope, will provide "buy in" from the other services and CINCs during development.

PROJECT MILESTONES

- **FIRST DRAFT**   **SEP 99**

- **SECOND DRAFT**   **FEB 00**

- **PRELIMINARY COORD**   **JUL 00**

- **FINAL COORD**   **SEP 00**

- **APPROVAL**   **NOV 00**

These are the project milestones delineated in the program directive.

ANNEX 7: Q & A FROM MOUT DOCTRINE PANEL

Question: There are a number of cities that have been abandoned along the Mississippi River because of constant flooding. These cities have been cleaned by the EPA. Fuel tanks have been pulled out of the ground, as have telephone wires and anything else that could contaminate the river. But the structures are still there. Can somebody look into possibly using those cities for urban training?

Unknown respondent: We are doing that.

Question: Several of you talked about urban inhabitants, but I didn't hear anybody mention civil affairs or psychological operations. Do any of you see that as being a critical part of the doctrine that we're working on?

Major Sumner: Let me take that one first. That was discussed in the program directive. It will be incorporated. We had deleted it as a stand-alone chapter because there is a CMO publication being developed as we speak.

Major Newell: The same is true of the handbook. We think that's a critical element and it's woven through the handbook. The Haiti case study is an example of good CA use and special operations in general. Grozny, in turn, is an example of poor use.

LTC Reardon: I will not discuss the topics as they pertain to the draft of FM 90-10 in the interest of time. They are woven throughout the portion we're going to have on civilian considerations.

Question: I got the impression from the briefings that the U.S. Army and Marine Corps may have slightly different approaches to MOUT.

LTC Reardon: In the Army and the emerging FM 100-5, *Operations,* there's a battlefield framework of offense, defense, stability, and support. Those will all be discussed in FM 90-10. In this briefing I keyed on what we felt was the most difficult aspect of MOUT, and that is conducting conventional offensive operations in urban environments. There'll be significant discussion on stability and support in the new FM 90-10, so we're also going to be touching on the low end of the conflict spectrum. There will be separate chapters for that discussion in the new manual.

Major Sumner: We have a separate chapter on OOTW in our service publication [MCWP 3-35.3], but most of the pub is focused on fighting in the city. We view that as being the hardest thing to get done. We're not looking at urban operations as a low-intensity activity.

Question: What are the specifics of map requirements, particularly with regard to scale?

Major Sumner: [Among several relevant factors], it depends on the city, quite frankly. You could have a very detailed map, but if it's out of date or if the city has suffered considerable damage, that may not help you much. You might have to depend on aerial photography.

John Allison, Marine Corps Warfighting Lab: You bring up a key point. One of the things I'll talk about tomorrow is what we've done with NRO and NIMA in regard to mapping and related issues. We have had some major challenges in determining the type of map products we need for the individual marine in order for him to leverage the technology extensions we're linking with close air support. It is a big challenge. Bringing it together with C4I and how you pass the shared tactical picture are some of the challenges that are harder yet.

Question: Can you give a sense of the differences in operations demanded when operating in dense urban environments versus the conduct of operations in environments characterized by urban sprawl?

Major Sumner: Anytime you get into a more vertical environment it's just going to eat up people that much faster. The service doctrine that we have talks about various types of urban terrain and how a city can be configured, but we don't go into detail on how fighting in a suburb differs from that in the city proper. It has to be based on the commander's estimate and his judgment.

Major Newell: In the handbook we make some general distinctions between different types of urban terrain, but a lot more work in that area needs to be done. The RAND Project Air Force guys are making some distinctions, I think. I think it's of interest that a lot of times during our planning we talk about worst-case scenarios for instance;

in our ongoing Joint Staff C4 study they consider downtown Manhattan. Even Mogadishu, with one- and two-story buildings primarily, was a significant challenge. We need to do more work in that area.

LTC Reardon: In the Army manual we discuss it in the IPB portion. We touch on it in terms of planning considerations and execution, but to echo my Marine Corps compatriot, we don't have different types of tactics for different types of urban environments.

Question: Is the doctrine addressing target designation for aircraft direct fires?

Major Sumner: We've used unit SOPs that specify coding buildings with letters and numbers. It is a very difficult problem if you are addressing CAS issues, regardless of whether you're talking fixed or rotary wing. You really need something, laser designators or something like that, to designate the target.

Question: We heard a lot of discussion about WMD concerns earlier. How are we handling them in the joint handbook and in the Army, Marine Corps, and joint MOUT doctrinal literature?

Major Sumner: I think we cover WMD in the chapter on consequence management. It's not being overlooked.

Major Newell: I don't think we brought it up in the handbook at all. We haven't dealt with WMD. It's obviously an important element. Maybe it's something we need to go back and look at putting in there.

LTC Reardon: We talked about it, and included cases involving inadvertent WMD-type events, like somebody accidentally taking out a chlorine storage facility with a Hellfire missile. It's part of our planning. We tell commanders to be very deliberate about the weapons they will employ when they go into a MOUT environment. The approach selected is also influenced by how much of the city we are willing to demolish and how much in the way of support responsibilities we are willing to accept for the population once we complete the operation.

INITIATIVES/TECHNOLOGY PANEL PRESENTATIONS

ANNEX 1: U.S. ARMY ACTD
MAJ Lee G. Offen, USA

MOUT ACTD
MILITARY OPERATIONS IN URBAN TERRAIN
ADVANCED CONCEPT TECHNOLOGY DEMONSTRATION

DBBL, Fort Benning, GA

- **Why a MOUT ACTD?**

- **Mission and Concept**

- **Requirements**

- **Candidates**

- **Program Overview**

- **Program/Experiment Methodology**

What I'm going to give you today is essentially an encapsulated information briefing on the Army MOUT ACTD. Just to recapitulate a little bit of what was said yesterday by Dr. Andrews, the objective of an ACTD is to transition items that have military utility into the hands of users.

MOUT ACTD

MILITARY OPERATIONS IN URBAN TERRAIN
ADVANCED CONCEPT TECHNOLOGY DEMONSTRATION

The most complex and resource intensive battlefield in the 21st Century

The most likely battlefield in the 21st Century

Potential adversaries have military parity with US forces in a MOUT environment

---

I don't even know that I want to talk about this slide after the past day and a half, but I will tell you that these are essentially the three premises upon which the conduct of the ACTD is based. We may have entered the information age, but that does not necessarily give us the advantage during MOUT that it gives us in other environments.

The four explosions are what we're all about. We seek to improve the operational capabilities available to soldiers and Marines. There are two pieces to this.

I spend a lot of time on the systems focus, those systems that give us the technological edge that will make our soldiers and Marines more survivable and lethal in a MOUT environment. There are tactics, techniques, and procedures, however, that need to go with those systems when we put them in the hands of the user. That is in the spectrum of battalion and below, all the way down to instruction of the individual soldier or Marine.

MOUT ACTD

MILITARY OPERATIONS IN URBAN TERRAIN
ADVANCED CONCEPT TECHNOLOGY DEMONSTRATION

IMPROVE

C⁴I

ENGAGEMENT

MOBILITY

FORCE PROTECTION

Provide technological dominance in MOUT for soldiers and marines

There are essentially four areas of improvement that the ACTD seeks to address: C4I, engagement (which as an old cavalryman I would describe as lethality), mobility, and force protection (which is really survivability).

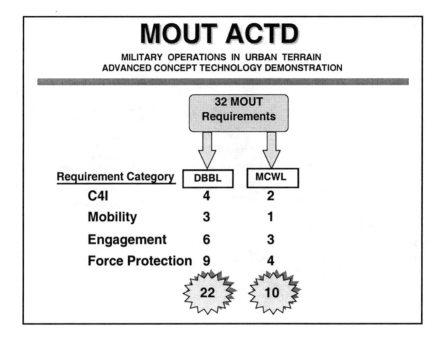

We have 32 requirements that we are looking at. They are based on 32 identified deficiencies. You see here how they are broken down. This is clear physical evidence that this is, in fact, a joint ACTD.

MOUT ACTD

MILITARY OPERATIONS IN URBAN TERRAIN
ADVANCED CONCEPT TECHNOLOGY DEMONSTRATION

Thru-wall scanning capability for
MOUT Squad Leader

Improved MOUT Individual
Mobility Tools

Physical Protection Kit (Eyes,
Joints, Medical, Hearing, etc.)

Non Lethal Munitions

Remote Intel
Collection Capability
for Battalions &
below

Enhanced Situational Awareness for
Individuals/Leaders in MOUT

Improved Squad Radio—
Effective in MOUT

Precise Position Location
inside of buildings

Combat ID System—
Effective in buildings

Improved All Environment Optics (Day,
Night, I2, Thermal, etc.)

Real time
Intel Down-
link to Map
Products

Increased Lethality/ Wpns
Capabilities for MOUT

These are some of the things that we're trying to provide for soldiers and Marines. We have done so with varying degrees of success. They include through-wall scanning, individual mobility tools, (mechanical breaching devices), and physical protection. I recently read about the catastrophic injuries that flying debris can cause during MOUT. So that's something we need to look at. There is also a nonlethal portion of the ACTD. Nonlethal systems are being looked at in the joint nonlethal group. You see the squad radio up there. Combat ID is still a challenge. Intel collection with UAVs and UGVs, and we've looked at some other things out there that are off the shelf.

MOUT ACTD

**MILITARY OPERATIONS IN URBAN TERRAIN
ADVANCED CONCEPT TECHNOLOGY DEMONSTRATION**

TECHNOLOGY CHALLENGES

- Small portable UAV and UGV suitable for employment at platoon/company capable of using interchangeable intelligence collecting payloads

- Soldier intercom able to operate in urban terrain with a "busy" electromagnetic spectrum

- Small hand held device able to see through walls

- Man-portable system to provide the ability to detect mines and booby traps in a MOUT environment

These are where we're finding our challenges. The UAV and UGV business is currently a cottage industry. There's an incredible variety of systems out there. We're still looking at those. Understand, however, when you talk about Marines and light infantry soldiers, they have to carry many of these systems themselves. That's a challenge when someone presents you with a 200-pound UGV. So portability is a big concern. It is progressing over time. Will it progress to the point where we come up with something that we'll be able to give to units starting in September 2000? That remains to be seen. We still are in the midst of experimentation even as I speak. We're going to have to come up with a soldier intercom that can be tailored to the available parts of the electromagnetic spectrum wherever soldiers or Marines are deployed. It also has to be powerful enough for people to communicate between buildings where there's no line of sight, but not so powerful that they become a target as a result of their transmissions. A small, hand-held device able to see through walls? Probably not in my lifetime. There's no question that some of these technologies will be available eventually. We just may be in a situation right now where we may have to accept partial solutions in order to achieve a degree of success.

I wanted to put up a slide that addresses our experimental timeline. The Marine Corps does something akin to this as well. We're on the verge of starting Army Experiment 6 next month. The Marines are soon starting Marine Corps Experiment 4 also. We're on parallel tracks as we look at our requirements. In July and September there will be joint experiments conducted that will provide an opportunity to look at the synergistic effect of all the winning systems together and how they contribute to the conduct of MOUT. We identify what requirements we're going to look at during the experiments. We look at the doctrinal piece, what the operators need to know in order to use the systems. We have contractors who develop a degree of familiarity with the systems in order to present observations prior to issuing them to the soldiers participating in the experiment. We then actually conduct the experiment and do our reports. OPTEC is the Operational Test Evaluation Command. It's the Army's operational test and evaluation agency. They help us in this process. It makes sense because if any of these ever become systems worthy of acquisition, they're going to be the guys looking at them.

MOUT ACTD

**MILITARY OPERATIONS IN URBAN TERRAIN
ADVANCED CONCEPT TECHNOLOGY DEMONSTRATION**

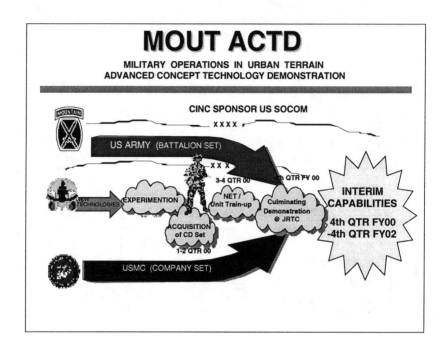

The Army and Marine Corps will continue to work together in the future in their joint efforts to better prepare men and women for future urban generations. Some of the near-term major events are shown here.

ANNEX 2: U.S. MARINE CORPS ACTD
LtCol John Allison, USMC

LtCol John Allison
Director MCWL Experiment Operations

I'm not going to talk to you about the Marine Corps ACTD program. It really is very similar to that of the Army, and Major Offen did a great job talking about the process. What I am going to talk to you about today is the Urban Warrior AWE [advanced warfighting experiment] and some of its technologies. One thing I'd like to say is that holding conferences such as this is absolutely the right thing to do. We need to talk about these issues, to bring them to the table, to get the people such as General Delk who came forward with his experiences in Los Angeles and the other folks who have come forward with their experiences. We need such forums so that we can make progress in dealing with the urban environment.

Since the collapse of the Soviet Union it has taken the services 20 years to really focus on military ops in the urban environment. We have seen how our missions are migrating to the urban environment, i.e., Somalia, Bosnia, Haiti, Kosovo, etc. We have always done basic MOUT training, but this has been done primarily at the company level and not in a combined arms manner. Our training facilities were just to small for MAGTF or brigade-level exercises. What we will not have the luxury to do in the future is spend 20 years educating and training our service members on MOUT and combat/HADR in the urban environment. We know that we are going to continue to send our men and women into harm's way in the urban environment. We want to give them the best training, the best tactics, the best procedures, and the best doctrine that money can buy for our force. We're doing it today. Just like LtCol Schattle said, we're putting people into the urban environment everyday.

The establishment of the Urban Warrior experimental area of operations (EAO) took a great deal of time and effort. One of the things I'm hearing here, and I'm very pleased to hear, is that the JRB has come forward with a recommendation to get a joint urban training center. When you try to do something of this nature, you have to coordinate training sites, medical support, security, police, firefighters, communications systems, range control, and all those types of things you take for granted on a military base. It was an incredible undertaking. It took a huge number of man-hours and a lot of cooperation with city agencies, local, state, and national government officials, and so on and so forth. But those are the types of things we need to do if we're going to put Marines and soldiers in harm's way and urban environments around the world.

I want to give you Professor Allison's history lesson, a little 101 about the Civil War. This is about the 1862 Battle of Fredricksburg. It seems that a Union commander by the name of Ambrose Burnside recommended a plan to take Richmond to President Lincoln, but what he'd have to do is go through Fredricksburg, an urban environment, to get there. The President told Burnside "I approve the plan. It will succeed if you move very rapidly, otherwise not." Well, on November 17th, Burnside got one of his divisions right outside of the town. At that point the Confederates only have a small force in the city. Then things start to happen: a little bit of bad weather, then the river's swollen, Burnside thinks he needs people, and so on and so forth. Eventually the Confederates and General Lee moved into the city. Almost a month later, the Union forces finally attack. By then, we're looking at 122,000 Union forces outside the city of

Fredricksburg, but now there are 78,000 Confederates there. Burnside finally does attack on the 13th of December. The Union forces cross the river and sack Fredricksburg. The citizens run in panic, and at that point the Confederates begin to establish defenses at a place called Marye's Heights in a sunken road. It is the high ground overlooking the city. The Union troops attack. There's a wall in front of the Confederates and they proceed to slaughter Union troops. Nine thousand casualties on the Union side, a little over one thousand on the Confederate side. That was urban warfare in 1862. Now what do you think any of the casualties on the Union side might have said about their training for urban warfare, about their tactics, procedures, and technologies? They'd probably have had a lot to say. That's what the War Fighting Lab is about. We're not about fixing the strategic problems that were talked about yesterday. We're not even about fixing the operational decisionmaking process that was talked about yesterday by several flag officers. We're about taking it down to the level of the individual who has to go into harm's way, because we know we're putting him there. So what I want to talk about today are some of the technologies that we have been looking into.

Urban Warrior AWE
After Action Review

". . . fighting through nine floors of the Oak
Knoll hospital was *really* hard. It opened my
eyes to what I have to face inside a modern
building with several floors full of bad guys."
Squad Leader, Charlie Co, BLT 1/5

Here's a quote from one of our squad leaders. During our Urban
Warrior experiment in Oakland, we found out that the casualty rate,
just as is in Chechnya, was still high. We have not solved this problem. But the actual experiment covered the whole spectrum of the
three-block war. To address this problem, the Warfighting Lab is
continuing to do urban experimentation with small-unit combined
arms tactics in Project Metropolis.

Overview

- Phase I: Monterey (13 March)
  - Naval Postgraduate School (NPGS)
  - Defense Language Institute (DLI)
- Phase II: Transition to Bay Area (14 March)
  - Moffett Federal Air Station
  - Concord Naval Weapons Station
- Phase III: SF Bay Area (15-18 March)
  - Embarcadero and Pier 35 in San Francisco
  - Oak Knoll Hospital in Oakland

This is just a quick breakdown on the dates. We found that hiring professional actors helped us quite a bit during Urban Warrior. They were more believable than Marines trying to play civilians. They did a superb job, even to the degree that a few of our coalition force members got hand-to-hand with a couple of them because they were playing their roles so well.

Urban Warrior Future MOUT Concepts

Urban Warrior experiments based on advanced concepts in seven areas:

- Command and Control
- Mobility/Countermobility
- Measured Firepower
- Survivability
- Adaptability
- Sustainability
- Awareness

Concept-based experimentation: if the concept's right, then let's go forward. Back in the 1920s and 1930s, as the Corps was experimenting with the amphibious assault concept, we were told it was folly. We said, "No, it can work." It just wasn't being done right. We took the concept and went forward. The Higgins Boat hadn't been made yet, but we knew the concept was right when we experimented with it. Yeah, sure, Marines fell out of wooden boats and these boats turned over. There were a lot of problems. But eventually we got the right technology because the concept was right. Well, that's what we're trying to do here.

Just a reminder: we're concept-based and we look at future MOUT concepts. These are the seven areas we'll talk about. I'll show you the technologies that go into them.

Dragon Fire

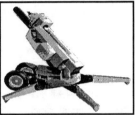

- Autonomous/minimally manned 120mm rifled mortar
- Designed to provide highly responsive battlefield support
- Receives digital target coordinates and automatically resolves fire solution
- Transportable via MV-22
- Emplacement within 5 minutes
- 8,300m range, 13,000m RAP

• Digital links and automated weapon systems demonstrated successfully in field events

• Requires additional development of associate systems—e.g. autoloader

• Needs to be employed in more tactical events to fully evaluate its contribution

This is the Dragon Fire. It is one technology that has a lot of promise right now. Our NATO allies like it. It's something you can set up in an urban environment. You don't have to have men to actually man it. And you can also put it into your system so that C4I can do a fire mission very rapidly, very quickly. The problem with it is that there is no auto loader. So after you fire a round, somebody's got to go back there and put a round in it. We don't want it to be that way, so we want to develop the system further.

Viper (PITS) Precision Intelligent Targeting System

- Uses GPS and laser range finder to compute ten-digit grid coordinate of target
- Lightweight binoculars enable ranging beyond 4 km
- Uses commercial camera batteries (binoculars) and AA batteries (PLGR). Precision Light Weight GPS Receiver
- Links to EUT for automated call for fire

• Well-suited lightweight, alternative to the target handoff systems being developed for observers

• Suffered from the same beam scattering problems of all laser devices in dense urban terrain

Something that we recently did during our AWE is link our end user terminal for automatic call for fire to a precision intelligent targeting system. We took it down into the urban canyon to see how well it would work. It had the same problem that any laser does: around all that steel, glass, and other things like smoke it was reflected or its range was reduced.

Tactical ATA (Airborne Target Acquisition)

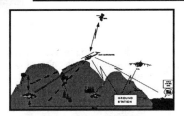

- Airborne sensor
- Quickly and accurately locates and identifies ground targets
- Rapidly transmits targeting data to direct support units
- Rapid call for fire/adjustment
- Automatic target handoff to weapons platforms
- Targeting accuracy: 20m Circular Error Probable @ 5 km

• Experiments limited to surrogate form (equipment mounted on a helicopter)

• Results very promising both in direct support of small units ashore and commands operating from a sea base

• Requires additional experiments with actual UAV

Airborne target acquisition system: this worked well. We had to use helicopter surrogates quite a bit because we couldn't fly actual UAVs over the civilian environment due to FAA and civilian safety concerns. It worked well and we think it has a lot of promise.

Combat UAV Target Locate and Strike System (CUTLASS)

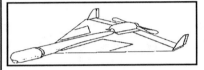

- Employed in surrogate form only (equipment mounted in a helicopter)
- Automatic target recognition software requires refinement for use in urban settings
- Tradeoff value as a weapons system requires more experimentation

– UAV designed to provide a picture of the battlespace via onboard sensors downlinked to a ground control station.

– UAV equipped with Automatic Target Recognition (ATR) payload resulting in a capability to "hunt" for targets contained within its database

– Battlespace observation

– Precision lethal fires

– BDA adjudication

The cutlass: this is a killer UAV. It can go in and find the target. Aviators like it because it has the potential to actually locate items like anti-aircraft and take them out. You can also use it for reconnaissance missions. We still want to do more experimentation to determine the tradeoff. What's the cost? Can we afford to have a system of this nature?

Boom Gun

- Crane that has remote control .50 caliber machine gun (with camera) at end of boom

- Uses same weapon mount as Mobile Counter Fire System

- Used for situation awareness and force protection

• Proved to be a very good idea
• Technology used was not sufficiently developed to ensure adequate reliability
• Requires more work on tactical employment to fully assess

Boom gun: this idea came from some Marines who suggested we look at mounting a sniper system remotely. We like to take ideas from the bottom up. Remember my example—the 9,000 casualties? The people on the receiving end might have some real good ideas. We like to listen to them. This is one and it does show promise in an urban environment. How do you see up and over things? Well, we mounted a camera and a gun on an elevated platform. Maybe it's not the right piece of kit yet, but the concept seems right.

Intrusion Detection System

– Three intrusion detection / perimeter security systems will be evaluated:

- Rapid Deployment Intrusion Detection System (RDIDS)

- MIL PAC 385B Relocatable Rapid Deployment Microwave Intrusion Link

- Video Motion Detection System (VMDS)

• Worked well during live force experiments

• Support units maintained perimeter security with minimum personnel

– Systems will be used to improve perimeter security and reduce the number of watchstanders.

– Improved force protection.

Nothing real sexy about this. It works, maybe we need to do better, make them smaller, have more of them. You can reduce the number of people that you need in certain soft-target areas which typically are where you have your combat support elements. You can monitor these systems and reduce the number of people who have to be assigned to security missions. In Mogadishu, a whole lot of Marines and soldiers were always pulling security. We had this big logistics dump down there at the airfield; in fact, there was a period when every coalition force was securing itself around the airfield. It was unbelievable. It got to a point where it was hard to put a plane in there because of everyone pulling security on the airfield. Maybe we can reduce the numbers we need for actions of this nature and then tie into another system that provides rapid reaction. Perhaps we can take the boom gun and use it for reconnaissance. We might well put a night-vision capability on it and so forth.

Force Protection Clothing

- Experimental Urban Camouflage Uniforms
- Reversible Urban Uniforms
- Thermal Thellie Suits
- Improved ability for marines to avoid detection on the urban battlefield
- Improved load bearing capability
- Improved ballistic protection for individuals

• Urban camouflage pattern proved to be very successful

• Improvements are indicated for protective accessories in order to enhance wearability

I said we do things that span the high end to the low end. Here you see a Marine with knee pads and elbow pads. We've already found out Marines don't like the elbow pads. They're too restrictive. They also think the knee pads need to be designed so as to be more like football pads, where you put the pads inside the uniform so that they're a little less restrictive. Those are the kinds of things we find out that then help these Marines as they're hitting concrete, steel, ladders, or whatever as they move through the urban environment. We found out that the uniform does have an IR capability that reduces the signature in an urban environment. That is another benefit that we're looking at with this type of stuff. One suggestion we got was to make the uniform reversible, maybe gray for urban environments and brown or green depending on where you're going.

Combat Decision Range

- Train NCO's in Combat decision making and leadership
- For Use by Every Infantry Regiment in the Marine Corps
- Operational March 1999
- 3 Man Mobile Demonstration Team
- Computer Based Training System
- Uses 7 Major conflict/combat Scenarios
- "Trains the Trainers"
- 3 Days of Instruction

• Extremely successful training means

• No hard data on performance improvement, but universally hailed by commanders and trainees as a desirable product

Something that has really been good is this combat decision range. We've run a combat squad leaders course that we've turned over to our Training and Education division. They are formalizing the course for use in the operating forces. We run squad leaders through a variety of scenarios, and we require them to use new technologies like squad radios and hand-held GPS. It's good stuff. We are looking at sending it over to Kosovo for our Marines on the ships there.

Auragen Generator

- The AuraGen G5000 is a 5Kw mobile generator that is mounted internal to a HMMWV.

- The G5000 operates at any engine RPM and generates 60 Hz AC power that can run the most sensitive electronic equipment.

• Technology appeared to work well

• Ultimate utility of using current TE vehicles as a substitute for dedicated electrical power devices requires more experimentation

- Reduces logistical footprint ashore by providing an organic power source and lighting capability.

The Auragen Generator is vehicle mounted. We are using it to potentially solve the battery problem that you have when Marines are using their end-user terminals.

Portable Reverse Osmosis Water Purification System (PROWPU)

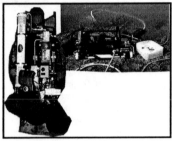

– Man portable system

– Capable of supporting 42-man infantry platoon (250 gallons)

• Capability judged to be a promising idea for the support of isolated small units

• Portability requires improvement if intended as a man-packed item—too heavy!

This is portable water purification. The need's there, but it's a little too heavy right now. We are currently working with the developer to reduce the weight.

Advanced Surgical Suite For Trauma Casualties (ASSTC)

– A rapidly deployable, modular structure that supports trauma management, resuscitative surgery, ancillary services, and temporary patient holding.

- Not employed tactically during UW

- Set up and layout appears to meet the requirement

- West coast units believe Mobile Expandable Container Configuration (MECC) may be a more promising technology for this capability due to its larger physical capacity

– Equipped with an environmental control unit and is capable of sustaining 25 trauma cases prior to resupply.

Advanced surgical suite, trauma casualties; we used this on the east coast. We had a mobile expandable container configuration on the west coast. The difference is (1) the expandable can be mounted into sea containers, moved on a trailer, and (2) it's self-contained so you can wipe the floor. The medical people like that because then you can sanitize it.

Scissors Truck

- Provides transportable above-ground sustainment or casualty extraction
- Commercially available
- Works up to 40 feet high

• Truck version only employed during east coast experiments with limited success

• Trailer version used during AWE with some utility

• Requires significant development

This is another one of those things that is not real sexy, but seems to have utility, at least in concept. It provides another way to move casualties out of a multiple-story building and things of that nature, including moving supplies.

MV-22 Compatible
Tactical Vehicle

- Assault variant provides a promising solution to the helo transportable vehicle requirement
- Met all mission requirements during AWE
- Logistics variant not used due to its size (not internally transportable)
- Requires additional experimentation to provide valid assessment of tactical utility

– Internal and external MV-22 compatible

– Highly agile, extremely stable on all terrain

– Can carry 3000-pound payload (Four personnel and equipment)

– 2 variants:

  • Light Strike Vehicle

  • Light Tactical Vehicle

This type of vehicle provided us with the right technology solution to our mobility and OMFTS challenges.

GATOR

- Very successful employment during east coast experiments

- COTS vehicle may require improvements for military employment

– Provides mobility on surface, subsurface, and structural layer of the urban battlespace

– Provides a resupply capability to the MAGTF

– Commercially available with proven logistic support

Gator: you may have seen this already on one of the previous slides. We use it for moving our supplies, logistics, for medevac, and so on.

Unmanned Watercraft
(Tactical Seadoo)

– Unmanned, remote controlled, high speed watercraft

– Delivers sustainment and is capable of emergency medevac

– Provides resupply without putting additional personnel or vehicles ashore

• Experimentation limited to demonstration of associated technology

• Requires significant additional work to assess tactical applications

– Craft has 6-inch draft and can operate autonomously in the open ocean or inland waterways via assigned GPS waypoints

This is one we've had success with. From 20 nautical miles off the coast we actually took a mannequin and moved it as a mock casualty back to the ship. It shows a lot of promise for a variety of missions and payloads.

Experimental Combat Operations Center (ECOC)

- Successful integration of experimental and legacy systems

- Overall operation was promising—requires additional development

- Testbed for tactics, techniques, and procedures and advanced Information Technologies

- Integrated applications that support fires, maneuver, aviation, logistics, and intelligence functions

- Computer-aided decision support

Here's our experimental combat operation center. A lot of what we do is C4I; we're looking now at how we do that from a forward base. We conduct over-the-horizon communications from the ship. Here is a picture of the ECOC in the sixth deck of the USS *Coronado*. We also put a smaller deployable ECOC forward in the littoral battle-space ashore.

Integrated Marine Multi-Agent Command and Control System (IMMACCS)

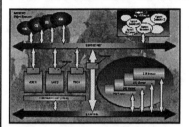

• As a proof of concept, worked extremely well

• Requires significant refinement to assess its full operational potential

The centerpiece of the experimental C2 systems is the Integrated Marine Multi-Agent Command and Control System (IMMACCS). IMMACCS attempts to demonstrate the utility of a comprehensive suite of automated decision support tools designed to interact as a whole. An innovative aspect of IMMACCS is the cutting-edge, multi-agent software which interacts with real-time information to provide the commander and his staff with useful "knowledge" about the battlespace.

IMMACCS: we've developed this in conjunction with several other agencies. The Joint Propulsion Laboratory with NASA is looking at this as a potential technology to help with the mission to Mars because one of the things that we do is have agents interact with one another in the battlespace. These agents actually can tell you something about the battlespace. Maybe there is a church on the other side of that hill where you thought you were going to fire an artillery mission. The agents can be set for an ROE alert and pick the fire mission up, alerting the commander/watch officer and shooter that this is an ROE problem.

Geospatial Information Database (GIDB)

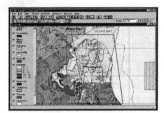

– Enables the MAGTF to access national geospatial databases and build digital mapping "views" needed in the ECOC and EUTs

– Totally digital architecture for delivering mission-specific geospatial data from the producer (NIMA) to the MAGTF and its subordinate elements

• Promising technology

• Successfully linked SPMAGTF(X) to national data bases and provided products that were translated into the IMMACCS object data base.

• Requires more experimentation to fully explore its potential

IMMACCS uses this. We need a digitized map system so that we can make sense of the symbols and graphics, and then we send it down to the individual squad leader who has an end-user terminal. We want to give him the appropriate information that the commander needs him to have for mission accomplishment.

GOSSIP (Ground Observation Special Support Intelligence Program)

- Provides radio identification of personnel placed on specific lists

- Based on "GangNet" software used by the LAPD and California Highway Patrol to track urban gangs

- Totally compatible with IMMACS and available to any EUT

- Achieved full integration into MAGTF C2 architecture

- Operators proficiency problems demonstrated the need to simplify or to devote more training time to equipped units

- Requires further experimentation in tactical scenarios

This one that could be used now during riots by organizations such as the LAPD and the California Highway Patrol.

Multi-lingual Interview System

– Hands-free operation using voice-recognition

– Resulting native-language speech can direct action or obtain a "yes or no" response

• Technical success during AWE both as a means to deal tactically with local populace and for enhancing medical care to foreign language speakers

• User interface aspects like the audio output and controls require refinement

This was a big success. We want to keep using it, especially for medical cases and as an interview system in instances where you grab somebody who you may not be able to initially identify as friend or foe.

Digital Automated Field Terminal

- Platoon commander's EUT suite
- Standard load-bearing vest modified to carry the following:
  - EUT Libretto 100 palm top computer
  - Wireless LAN 2 (Antenna on his right shoulder)
  - Differential GPS (Antenna on his left shoulder)
- Total weight is less than 12 lbs
- Provides individual instrumentation package

• Limited use of this and the individual EUT during East Coast experiments

• Limited technical success achieved during AWE—full digital connectivity with IMMACS

• Systems improvements and additional experimentation are needed to fully develop this capability and assess its impact

This isn't what we want our Marine of the future to look like, but that's the individual GPS radio system. He's also got another radio system, his end-user terminal. Of course, if he fell forward to dodge a bullet, he'd crack that thing. This isn't what we want. Again, it is the concept that is key; we're going to move forward with this. We're getting great ideas on how to change what this looks like.

Tactical Instrumentation

- EUT and IGRS provided combined instrumentation

- Technology proved promising: worked better during later stages of the AWE once user procedures were refined

- Additional functionality, e.g., life status, would greatly enhance tactical utility

- Significant enhancement to training/analysis efforts

– Real-time continual tracking of all personnel

– Integrated into existing C4I systems

– Indoor/outdoor

– Potential operational application

Tactical instrumentation: we were able to instrument the buildings and Oakland Hospital during Urban Warrior so that we could see where everyone was. Individual GPS radio systems showed where the friendly, enemy, and role players were. We had boxes inside the buildings. These boxes sent signals to show people moving because GPS doesn't work inside buildings. We think there is a possible tactical use for these boxes. As you go in and clear a building, you could leave the boxes behind and monitor them; if someone went in and broke its wire, you know that you have an intruder.

Joint Conflict and Combat Simulation

• Integration of JCATS and C4I system during AWE permitted expanded CTP
• Interior building instrumentation used to track combatants inside buildings
• JCATS generated enhanced operational level play by introducing additional constructive forces into the scenario
• JCATS provided computer-based resolution of indirect fires to live forces

– Newest USMC high-resolution model

– Provides details of urban terrain

– Level of resolution well suited to supporting MCWL experiments

– Planned technical integration with instrumentation system

Here's a picture of what we saw in JCATS: the red, the blue, and the neutrals. It gets updated about every 2 minutes. You could actually see individuals moving across the screen.

Aviation LTA
9 – 16 April 99

Yodaville

- Laser Guided Training Round
- Maverick KEO PGM
- Inert TOW
- Rapid Targeting System
- Tactical Airborne Target Acquisition System (H-1 surrogate)
- CUTLASS Lethal UAV (H-1 surrogate)
- Dragon Fire 120mm Mortar
- Dragon Drone UAV
- Pioneer UAV
- K-Max Helicopter
- LADAR (Laser Radar)

We talked to the aviation somewhat already. We could not do a lot of fixed-wing aviation in Oakland, so what we did is put it all into our Yuma exercise for June 1999.

Summary

- MCWL analysts will reconstruct the experiment data to glean information and develop knowledge
  - analysts accompanied troops in battlespace
  - subject matter experts were in every part of the battlespace
  - debriefing of participants was done on-site immediately following each evolution of the Experiment Plan
- Report due in June 1999

This gives you an idea of what we're going to be doing and what we have already begun. The final report will be due in June 1999.

Summary

Urban combat problem is not solved

- Physical and psychological challenges of modern, steel and glass highrise buildings require special—and repeated—training
  - not easy to apply existing and emerging TTPs without focused training
- Use X-Files to get word to operating forces
  - as information is inserted into the Combat Development System

I'll be honest with you. We haven't solved the problems associated with urban combat. That's a bloody one; it's hard to solve. If it were easy we'd probably already have done it. We're going to keep working it.

Marine 2010

Heads Up Display
- Situational Awareness Module
- End User Terminal
- Comm Image Receiver
- Weapon Reticle & Status Display
- Personal Health Status
- GPS Readout
- NBC Warning System

Face Shield
- Half Face Shield for Normal Ops
- Full Face Shield for NBC Protection
- Day/Night Fusion All Weather Imagery

Integrated Individual Combat Weapon
- Adjustable Lethality Weapon
- Automatic Fire Control System
- Automatic Identification System
  – Friend or Foe
- Laser Range Finder
- Laser Designator/Spotter

Integrated Helmet Assembly
- Integrated Combat Identification
- Comm Suite (Image/Video/Voice)
- Micro Battery Pack

Integrated Combat Utility Uniform
- Adjustable Camouflage Pattern
- Reactive Body Armor
- Integrated Load Bearing Equipment
- Medical Monitors
- Environmental Controls

Situational Awareness Control Panel
- Control for End User Terminal and GPS

Integrated Generator
- Heel Compression Generator

This is what it's all about. We want to prepare Marines, soldiers, airmen, or sailors to do the mission that their country's going to give them.

ANNEX 3: COMMUNICATIONS IN URBAN ENVIRONMENTS
Mr. Sean J. A. Edwards, RAND

Tactical Communications and Military Operations on Urbanized Terrain (MOUT)

Sean Edwards
April 1999

My name is Sean Edwards and I am a doctoral fellow at the RAND Graduate School of Policy Studies. My brief presentation this afternoon is on tactical communications in MOUT. I want to do this from a policy perspective. Although I touch upon technical terms and concepts, I leave most of the technical subject matter for the communication engineers among you.

By the way, my own experience with communications is limited to my fond memories of humping a AN/PRC-77 around for my company commander when I was a light infantryman in the Army. I think I speak for all infantrymen when I say *size does matter*.

Objectives

- **Outline the communications challenge in MOUT**

- **Note possible workarounds and solutions**

In the next ten minutes I have two objectives: to outline the basic challenge inherent in wireless communications for urban areas and to note ways to alleviate the problem.

First, let's look at the demand side of the communications challenge.

Challenges on the Demand Side

- **Looking at situational awareness as a demand problem**
- **Communication systems must be managed as a scarce physical resource**
  - **Information quantity—what is the right balance?**
  - **Information type —text, voice, still imagery, video?**
  - **Information destination—who needs it?**
  - **Information timeliness—real time or not?**

As you know, the Army and the Marine Corps are seeking to provide situational awareness for their combat formations. The problem is, this may not be possible in difficult environments such as sewers or deep inside multi-story structures. Because of this, we need to approach the problem from the demand perspective as well as the supply perspective. Think of it this way: communications resources must supply the information and data that warfighters demand, while that demand for information may sometimes need to be adjusted to the technological limitations of supply. Demand can be adjusted in several ways:

- **Information quantity:** How much is enough? Do you need location *and* status? A balance should be found between providing the soldier too much information—information overload—and too little information, which would fail to exploit the ongoing information revolution.

- **Information type:** Does the soldier need text? Voice? Still imagery? Video?

- **Information destination:** Who needs it? Not every command echelon has the same information demands. Whereas a battalion commander may find a real-time video teleconference with his company commanders a useful capability, a fire team leader in charge of clearing the floor of a building probably only needs voice and positional information.

- **Information timeliness:** Does the information really need to be real time?

Challenges on the Supply Side

- **Propagation**
  - Radio propagation problematic because of path loss and fading
- **Hardware limitations for man-portable radios (size, weight, power)**
- **Communications security**
  - Detection
  - Exploitation
  - Interception
  - Resistance to Jamming

We talked about demand; now let's talk about major challenges on the supply side.

First, in an urban environment, wireless communication suffers from path loss and fading. Transmitted signals are blocked, reflected, refracted, and defracted as they travel through and around buildings, walls, and floors. Received signal strength drops, which also lowers throughput.

Second, MOUT is largely an infantry fight, so most radios must be man-portable. Hardware size, weight, and power (read batteries) are a challenge.

Third, communications security and resistance to jamming continue to be major concerns, especially when one discusses the possibilities of commercial-off-the-shelf (COTS) systems. Warfighters are concerned with:

- Low probability of *detection* (LPD), which is keeping the enemy from knowing that friendly units are transmitting.

- Low probability of *exploitation* (LPE), which is keeping the enemy from fixing friendly unit positions.
- Low probability of *interception* (LPI), which is keeping the enemy from knowing what friendly units are sending.
- Resistance to jamming.

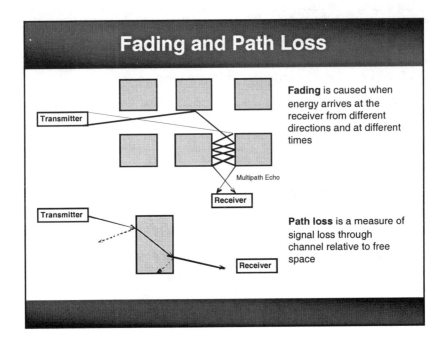

I already mentioned that fading and path loss are the most difficult propagation problems.

Fading, which is variation in received signal strength, occurs for several reasons. Fading (multipath) occurs when direct waves and reflected waves from the same signal arrive at the receiver from different paths and at different times (out of phase) and are subjected to destructive interference. Basically, fading occurs when radio waves from the same signal reflect off of obstructions and interfere with each other.

Path loss occurs when radio signals are attenuated as they pass through or around walls, buildings, and objects in the channel. Energy is absorbed or reflected as it passes through objects. Path loss is a term used to quantify the difference (in dB) between transmitted power and received power.

For cellular frequencies, private firms have conducted field experiments to measure path loss. This kind of information is usually proprietary, but some studies have demonstrated the relative path losses for various environments. For example, building construction materials, building age, wall locations, ceiling heights, and the like all affect path loss. Major obstacles include steel slabs, metallic pipes, and ventilation ducts. Concrete is worse than brick, and brick is worse than limestone. Floor level also matters. For example, middle floors are the best when building penetration loss is the concern and other tall buildings are adjacent. More windows help radio propagation.

Wireless Options

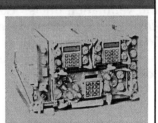

SINCGARS

- **LOS radios**

- **LOS microwave radios**

- **Cellular phones (terrestrial)**

- **Satellite-based phones**

- **Non-LOS Packet radios**

Given these challenges, what are the wireless options available?

SINCGARS is the current tactical line-of-sight (LOS) radio. It is already a part of the Tactical Internet, has good security, and has a range of 5–10 kilometers for man-portable sets.

LOS microwave radios have a higher data rate, but the antennas are too bulky for dismounted deployment.

Cellular phones and satellite-based PCS phones are usually seen as an attractive option. The Chechens used them in Grozny and the Somalis used them in Mogadishu. Satellite phones offer a mobile user the option to uplink to a satellite whenever a ground base station is unavailable. There are several problems, however:

- Cellular telephone systems require a fixed infrastructure of base stations and land lines which could be vulnerable in a firefight;

- Neither cellular nor satellite-based PCS phones have high data rates;

- They are to some degree incompatible with existing military systems; and

- Most important, both types of phones have poor security and are easily jammed.

Since commercial firms try to maximize profit, they use the most economically efficient waveforms available and do not worry about jamming. Waveforms with high anti-jam and LPD characteristics are to some degree incompatible with high spectral efficiency. They end up sacrificing security for greater spectral efficiency (bits per second per hertz).

The packet radio is a promising alternative. A network of packet-switching radios can exchange information in a store-and-forward fashion, so that the source and destination radios do not need to have unbroken LOS (unlike the systems described above). There is currently no packet radio available, but one may be available in the near future.

Objectives

- **Outline the communications challenge in MOUT**

- **Note possible workarounds and solutions**

Now that I have outlined the basic challenge, I want to suggest some doctrinal and technological options that might help alleviate the communications problem.

Workarounds and Possible Solutions?

- **New software radio**
- **Retransmitter concepts**
- **Ultra-wideband signaling**
- **Antenna option**
- **Adjust tactics**

The first four options are technologies that will help supply more information to the warfighter. These include the software radio, retransmitters, ultra-wideband signaling, and array antennas. The fifth suggestion is doctrinal: adjust tactics to lower the demand for information.

Software Radio

- **Packet switching, non-LOS**
- **Optimizes frequency, bandwidth, modulation, waveform for a given signal environment**
- **"Multiple personalities"**
- **9600 bps rate**
- **Under development by DARPA**

The software radio (also called wideband or transforming radios), is a packet switching, non-LOS radio that uses software applications to perform some of the major communications functions that analog components do in current radios.

Software radios can be thought of as having "multiple personalities" because they will be capable of jumping between frequency bands as necessary to maximize performance in restrictive environments. A smart version of the radio will optimize modulation, frequency, and power level to best suit the environment and desired security. The same radio might have three primary bands of operation: VHF band (30–87 Mhz) and two UHF bands (for example, 225–450 Mhz and 450 Mhz–2 GHz).

For example, if a soldier finds himself in a room, surrounded by wooden walls containing chicken wire, the soldier can switch from VHF to a higher-frequency UHF signal with shorter wavelengths capable of squeezing through the gaps in the wire. If he wants to go outside, he can switch back to a VHF signal that will propagate more effectively around buildings in the urban canyon.

The new software radio will be compatible with new waveforms (modulation and error-control coding) that haven't been invented yet.

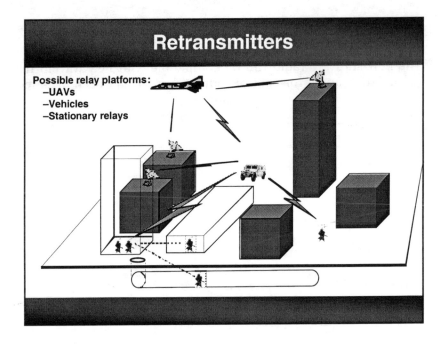

Retransmitters, or relays, can add communication nodes to a radio net. Relays can be mounted on buildings, aircraft, UAVs, vehicles, or dismounted soldiers to establish communication between units that are not necessarily in LOS. It may even be feasible to air deploy relays to building tops.

DARPA is investigating a drop off attritable relay compatible with SINCGARS and the future software radio.

Other Technological Fixes

Ultra-wideband signaling
- More covert because the signal is below noise floor
- Time-modulated waveform reduces multipath fading
- Resists jamming

Array Antennas
- Reduces fading

Ultra-wideband signaling is a promising new technique that is more covert, hard to jam, and fades less.

Ultra-wideband (UWB) signals are more covert than narrowband signals because UWB signals are spread across a larger band of frequencies than is required for normal transmission. As a result, the average power or amplitude in at any given frequency is indistinguishable from background noise.

Ultra-wideband signals also do not fade as much because they are time-modulated rather than amplitude or frequency modulated. UWB radio pulses are so short—about 40 million pulses a second—that the reflected versions of the signal do not cause destructive interference (fading).

The wide operating bands of UWB systems make it difficult for jammers to distribute enough energy across all the used frequencies (up to several Ghz in some cases).

Array antennas use multiple antenna elements at the receiver end (with a separation greater than or equal to half the signal wavelength), so that several separate, independently fading signal paths are established between the transmitter and receiver. The received signals can be combined for a stronger signal that reduces fading.

My last suggestion is to lower the demand for information by adjusting tactics, especially in very difficult electromagnetic environments with high path loss and fading characteristics, such as sewers or large multi-story buildings built of steel and reinforced concrete. In areas where wireless radio cannot operate, what you might call "communication dead zones," tactics may need to be based on zero situational awareness.

Perhaps a more realistic goal in the immediate future is to achieve between-building rather than in-building situational awareness. By between-building situational awareness I mean letting the warfighter know what is going on building-to-building, street-to-street, and block-to-block, not necessarily what is going on deep inside structures. I have displayed the Mogadishu example to make the point that between-building situational awareness can still save many lives. The Rangers and Delta Force commandos who fought in Mogadishu in October 1993 could have used between-building more than in-building situational awareness. Most U.S. casualties occurred during movement down streets and between buildings. They

actually did little room-to-room clearing or fighting inside very restrictive environments like the target building indicated above.

Now, we should keep in mind that there will be communication dead
zones in the future—individual infantrymen, fire teams, and squads
will continue to be cut off from their radio nets for one reason or another. General Scales mentioned that our enemies will adapt. One
thing they might try is to use radio frequency (RF) bombs. RF bombs
are grenades small enough to be hand tossed that set off a mini-
electromagnetic pulse that can fry communications gear, at least
temporarily. "Traditional" signaling techniques and tactics based on
limited or zero situational awareness, essentially the tactics we use
today, will continue to be relevant for MOUT.

Communication limitations have changed tactics before. Along the
western front in World War I, commanders lost contact with their
assault troops when they crossed into "no man's land" because the
field telephone system of the day was vulnerable to artillery fire. As a
result, when subordinate units went "over the top" they entered a
black hole from which no further communication with higher levels
of command was possible (except by runner). No communication
meant that all coordination between the rolling artillery barrages and
the advancing infantry had to be preplanned. All too often the
infantry got bogged down for some reason during their advance and
ended up watching with helpless horror as their protective artillery
barrage "walked off." Tactics were adjusted to use Very pistols as a
signal for artillery support after telephone communication was lost.
The history of command in war (and the search for situational
awareness) essentially consists of an endless quest for certainty—
certainty about the disposition and the intentions of the enemy and
your own forces. But as Clausewitz himself explained, uncertainty
and "friction" define the very nature of war.

Conclusion and Contact Information

Difficult communication challenges must be met with a mix of tactical and technological fixes

Contact information:

Sean Edwards
RAND
1700 Main St., PO Box 2138
Santa Monica, CA 90407-2138
PH # (310) 393 0411
Sean_edwards@rand.org

In conclusion, there is hope in some of the near-term technology initiatives, but our soldiers should continue to train for MOUT with tactics based on limited situational awareness.

My contact information is listed here; I look forward to your suggestions or questions if you want to contact me in the future.

Backup Slides

A Combination of Data and Voice Is Needed

Findings from the Marine Corps Warfighting Lab

- **Data is most useful:**
  - For predefined information (e.g., position reports)
  - For routine information (e.g., status reports)
  - For numerical data (e.g., target coordinates)
  - For rapid ("burst") transmission of high-volume information

- **Voice is most useful:**
  - When on the move
  - In high-tempo operations
  - For personal communications (e.g., hearing the stress in a subordinate's voice)

Field tests are beginning to answer these types of questions. Exercises by the Army and Marine Corps provide a starting point for determining warfighter needs. The chart at right is an example of how data requirements differ between unit types. Combat service support units, for example, require more data communication than voice communication.

ANNEX 4: URBAN OPERATIONS AND
THE ARMY AFTER NEXT
LTC Robert F. Hahn, III, USA

THE ARMY AFTER NEXT (AAN)
URBAN WARFARE PROJECT

RAND MOUT Conference

13-14 April 1999

LTC Robert F. Hahn II

Futures Directorate, Deputy Chief of Staff for Doctrine
United States Army Training and Doctrine Command

My name is Lieutenant Colonel Bob Hahn. I am a member of the Army After Next Project at the U.S. Army Training and Doctrine Command (TRADOC). My primary areas of responsibility within AAN are geopolitics, economics, and coalition operations. For the past two years, I've also had the opportunity to serve as the Director of the Army After Next (AAN) Urban Warfare Project. Over the next few minutes, I will provide you with a brief overview of some of the work we have been doing in the area of future urban operations. I intend to address the purpose and organization of the Urban Warfare Project, the operational concepts we have developed for use in the AAN series of war games, and a brief description of some of the weapon systems we have been looking at for use in the urban fight in the 2020 time frame.

Project AAN Mission Statement

Conduct broad studies of warfare to about the year 2025 to frame issues vital to the development of the U.S. Army after about 2010, and provide those issues to senior Army leadership in a format suitable for integration into TRADOC combat development programs.

I realize some of you are not very familiar with the Army After Next project. We were created by General Dennis Reimer, the Chief of Staff of the Army (CSA), in early 1996. The purpose of the organization is to provide an institutional process through which the Army can examine warfare in the mid- and far-term future, which for us represents the timeframe from about 2010 to 2025. We are organized to conduct our examination of the future along four major axes: geopolitics, military art, technology, and human/organizational behavior. Each year, with the help of a large number of franchise organizations, we conduct a series of studies, war games, and seminars which we then use to produce an annual final report for the CSA. The 1998 Annual Report was released in December and is entitled "Knowledge and Speed: Battle Force and the U.S. Army of 2025." It is available in both printed form and from the AAN homepage. We also provide input directly into the R&D and combat development community through the Office of the Deputy Chief of Staff for Combat Developments at TRADOC Headquarters.

Army After Next Urban Warfare Project

PURPOSE:

To examine the conduct of military operations in and around urban and complex terrain in the 2025 time frame with an emphasis on future warfare at the operational level.

 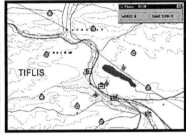

The early AAN tactical war games we conducted suggested that many of the concepts, organizations, and weapon systems we were examining did not work as well in highly urbanized areas as they did in more open terrain. In November 1997, we created the AAN Urban Warfare Project to study urban operations more closely. At that time we chose to focus primarily on the operational level of war. We have since expanded the purpose of the Urban Warfare Project to include an in-depth look at operations at the tactical level of war, down to and including squad and individual soldier activities. Most of our study of the tactical level takes place within the context of the Urban Integrated Idea Team (Urban IIT). I will speak more about the Urban IIT in a few minutes.

Definitions

- **Military Operations on Urbanized Terrain (MOUT):**
  All military actions planned and conducted on a topographical
  complex and its adjacent natural terrain where man-made
  construction is the dominant feature. (FM 101-5-1, 30 Sep
  1997)

- **Complex Terrain:** Any combination of natural or
  man-made terrain, urban infrastructure, and civilian population
  that negates or restricts the application of advanced
  technology systems and severely complicates maneuver,
  interdiction, and battlespace awareness during military
  operations. (nondoctrinal term)

As you have probably noticed, I used the term "complex terrain" on
the previous slide rather than the more traditional term "MOUT."
Complex terrain is a nondoctrinal term we created at the TRADOC
Analysis Center (TRAC) at Fort Leavenworth to better capture the
entirety of the challenge presented by the large urban expanses we
expect to encounter during warfare in the 2020 time frame. While
primarily referring to the urban environment, complex terrain can
also include other forms of restrictive terrain such as extensive, steep
mountain ranges. One of the critical components of complex terrain
tends to be the presence of a substantial civilian population. The
large noncombatant populations and urban infrastructure that we
expect to encounter during operations in complex terrain greatly
complicate the operational, logistical, and technical challenges being
faced by the Joint Force Commander. In many ways, urban operations will be fundamentally different from military operations on
open terrain.

Army After Next Urban Warfare Project

**AAN Urban
Warfare Project**

**Future Urban and Complex
Terrain Operations
Franchise**

Fort Benning, Georgia

**Urban Integrated
Idea Team (IIT)**

**Army Research
Lab**

The AAN Urban Warfare Project actually consists of three mutually supporting organizations. The core of the effort is located within DCSDOC, TRADOC at Fort Monroe and includes the primary AAN staff in the Futures Directorate as well as the near- and mid-term urban operations experts in the Joint and Army Doctrine Directorate (JADD). JADD serves as our link to the J-8 Urban Working Group and to the doctrine writers at Fort Leavenworth.

In general terms, AAN franchises provide an in-depth look at specific areas in support of the larger AAN effort. We currently have franchises that examine space and missile defense, special operations forces, information operations, force projection, sustainment, homeland defense, medical support, and urban operations.

The Future Urban and Complex Terrain Operations Franchise is located at Fort Benning, Georgia. The Urban Franchise is an extension of the recently created Combined Arms MOUT Task Force and is focused on developing concepts and organizations for brigade-level and below for future urban operations.

The final component of the Urban Warfare Project is the Urban IIT. Co-sponsored by AAN and the Army Research Lab, the Urban IIT brings together operators and technologists to conduct an integrated examination of AAN equipment, organizational structures, and operational concepts. This year, the Urban IIT spent a great deal of time developing the Future Soldier System with an eye on how that system could be used during urban operations.

AAN Urban Warfare Project

FY 99 RESEARCH APPROACH

- Urban Integrated Idea Team (IIT) (Nov 98-Jun 99)
- ALSA Aviation Opns on Urban Terrain Joint Working Group (19-22 Jan 99)
- 1999 AAN Force Projection Game (1-5 Feb 99)
- SOF Military Operations in Urban Terrain (MOUT) Wargame (8-11 Feb 99)
- 1999 National Security Seminar/Campaign Planning Workshop (14-19 Mar 99)
- 1999 Spring Wargame (26-30 Apr 99)
- Future Joint MOUT Initiatives (J8 Urban Working Group)
  - Joint MOUT handbook
  - Joint Concept
  - JP 3-06
- Joint Conflict and Tactical Simulation (JCATS) test
- FM 90-10, Military Operations on Urbanized Terrain
- ACOM Joint Concepts and Experimentation effort

This slide lists the major activities and events associated with the AAN Urban Warfare Project during fiscal year 1999. As you can see, the AAN Urban Warfare Project has sought to participate in most of the major activities mentioned by other participants in this conference. The remaining projects we still have ongoing for this year are the AAN Spring War Game and tactical excursions, the writing of the Urban IIT final report, and possibly a test of the Joint Conflict and Tactical Simulation (JCATS). We have been working closely with the program manager for JCATS on developing ways to integrate the capability JCATS provides into our existing suite of AAN war game models and simulations.

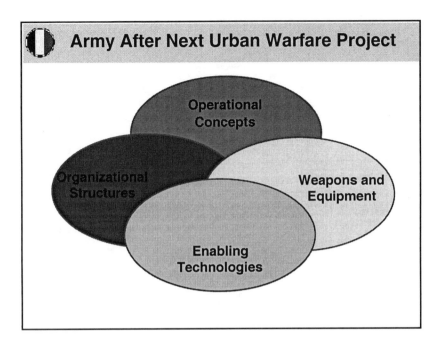

This slide should help you better understand how we approached the challenge of AAN-era urban operations this year. First, we produced a "Future Joint Concept for Urban Operations" that we have used as the organizing framework for the rest of our study effort. It is based on an overarching AAN concept that we call "Advanced Full Dimensional Operations (AFDO)." We then built a specific organizational structure, the "Light Motorized Battle Force (LMBF)," that we attempted to optimize for urban operations. As do most of the other AAN battle forces, the LMBF includes a combination of aviation, motorized infantry, and fire support assets. The difference is that we sought to combine them in a ratio that we felt was most applicable for urban operations of the kind our operational concept requires. During the upcoming AAN Spring War Game, we will examine the applicability and effectiveness of some of the specific weapons and equipment developed by the Urban IIT. Finally, based on what we learn during the war game and some tactical excursions we are going to conduct later this summer, we hope to be able to identify some key enabling technologies that we will include in the Urban IIT final report this fall.

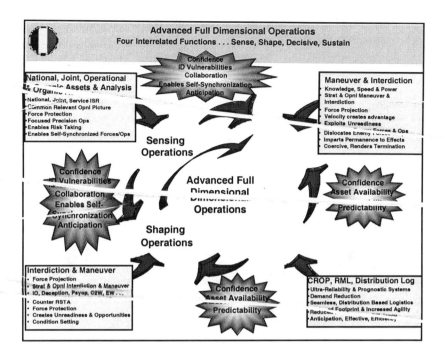

I am going to speak very briefly now about the four areas I have just addressed: operational concepts, organizational structures, weapons and equipment, and enabling technologies.

As I mentioned, the concept for urban operations that we developed is based on the AAN concept called Advanced Full Dimensional Operations (AFDO). This slide outlines the four major interrelated functions within AFDO: sensing operations, shaping operations, decisive operations, and sustaining operations. AFDO is our attempt to describe the rapid, simultaneous, continuous, and dynamic application of our integrated joint military capability in such a way as to cause the enemy to concede, disintegrate, or face certain failure. AFDO are made possible through a combination of knowledge, speed, and power. Knowledge is based on a combination of understanding of the environment and opponent before the conflict and the clear, integrated picture of the situation during conflict. Speed is achieved through a combination of knowledge and the physical agility AAN forces will possess. Power is the synergistic momentum produced by knowledge and speed coupled with the ability to generate sustained lethality.

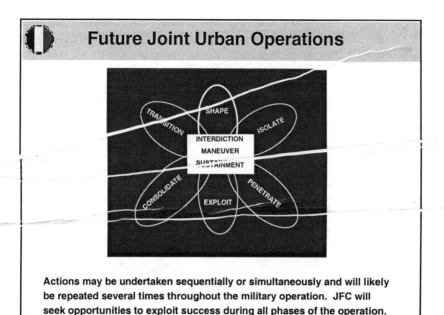

Future Joint Urban Operations

Actions may be undertaken sequentially or simultaneously and will likely be repeated several times throughout the military operation. JFC will seek opportunities to exploit success during all phases of the operation.

In addition to AFDU, we also based much of our Concept for Future Joint Urban Operations on the Operational Concept for Joint Urban Operations being developed within the J-8 Urban Working group. Our concept includes six interrelated activities that will likely be conducted simultaneously by the Joint Force Commander. These actions are shape, isolate, penetrate, exploit, consolidate, and transition.

Shaping actions are best described as employing key joint force capabilities to limit the ability of the enemy to use complex terrain to lessen or neutralize our operational advantage in speed, battlespace situational awareness, and precision munitions.

Isolation requires the identification and control of the movement of personnel, equipment, and information into, out of, and within the urban area.

Penetration includes all efforts to gain initial access to the urban area and to establish a force presence within the urban battlespace.

Exploitation is the application of maneuver and interdiction to destroy the ability of the enemy to conduct coordinated military operations within the urban area.

Consolidation actions are those taken to strengthen existing control over the urban area and to prepare for future operations.

Finally, transition is the transfer of routine control over the urban area to local security forces and civilian authorities.

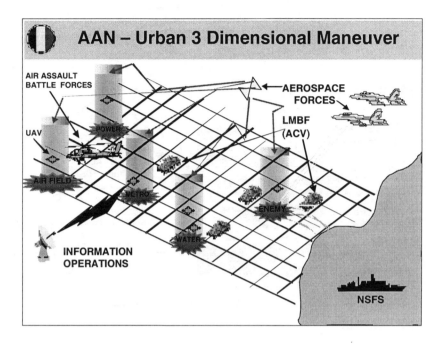

This slide is designed to depict what it looks like when we put all of the various pieces together using a combination of aerospace, naval, and land forces. The joint force is conducting what we refer to as three-dimensional urban nodal operations. Rather than trying to seize a limited foothold in the city and then expand control building by building, block by block, the force is simultaneously attacking critical infrastructure nodes and enemy positions throughout the urban battlespace using multiple forms of interdiction and maneuver. Many of these nodes would be the responsibility of relatively small units that will have extensive robotic and "reachback" precision fires capabilities. The shock produced by this operation would likely cause the enemy forces within the city to disintegrate, thus creating a situation where additional U.S. or coalition forces could be safely brought in, if necessary, to facilitate continued control over the city and establish the conditions required for consolidation and transition.

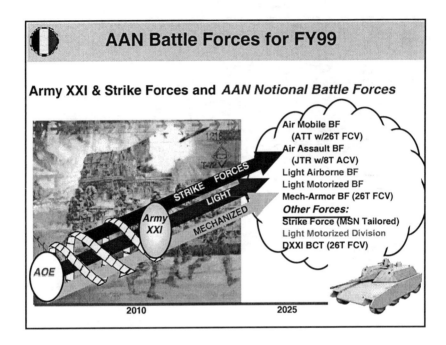

The Light Motorized Battle Force depicted in the previous slide is just one of five types of battle forces that we have developed for use in AAN war games. As you can see in this slide, we have battle forces built primarily around two different types of vehicle: a 26-ton Future Combat Vehicle (FCV) and an 8-ton Advanced Combat Vehicle. We also use various mission-tailored strike forces and other types of Army XXI forces during our war games. We have a complete description of our battle force designs, organizational structures, and operational concepts available on our Web site.

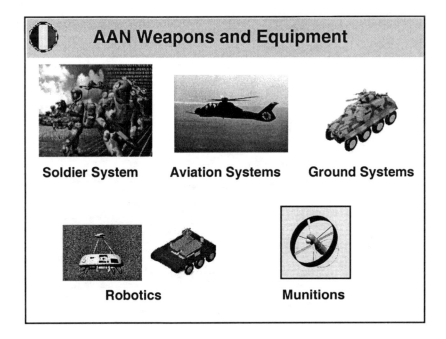

The primary focus of the Urban IIT this year has been to develop the various components of the Future Soldier System. We have also looked at the challenges of using the aviation and ground systems resident in AAN battle forces within the context of urban operations. These investigations have led us to conclude that much of our ability to successfully conduct small-unit urban operations in the future will require an extensive suite of unmanned systems and small, brilliant precision munitions. We believe that a small unit (company or platoon) equipped with the right combination of robotic systems and reachback fires will be able to control areas of a city that now require a very substantial commitment of traditional forces.

The Urban IIT plans to continue its examination of small-unit operations throughout fiscal year 2000 with an emphasis on the synthesis of manned and unmanned systems at the company, platoon, squad, and individual level. We will also be seeking to identify the specific unmanned systems and munitions technologies that need additional research and development funding in order to produce the capabilities we believe must be resident in the force for future urban operations.

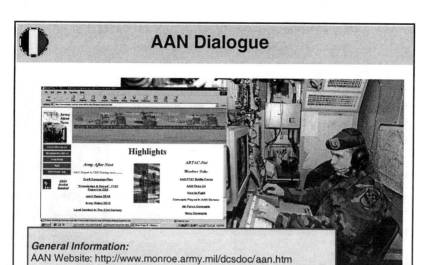

This has been just a very brief overview of the Army After Next Urban Warfare Project. If you are interested in looking at Army After Next in more depth, please feel free to visit our Web site at the address listed on this slide. The site contains the CSA annual report, all of our current briefings, and specific information about the efforts of the franchises and the integrated idea teams we sponsor. Thank you for your attention. I'll be glad to answer any questions you may have.

ANNEX 5: DECEPTION IN THE CITY
Mr. Scott A. Gerwehr, RAND

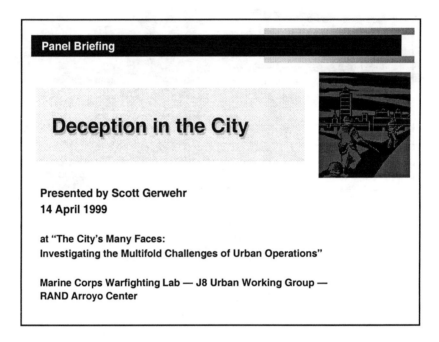

We've been discussing various important components of the urban operations challenge (communications, medical issues, etc.). An area that is perhaps underinvestigated—but has potentially high payoff on both offense and defense across the spectrum of intensities—is deception. The RAND Arroyo Center MOUT project has been studying the topic, and we present here a glimpse at our ongoing work.

Military Deception - US Joint Doctrinal Definition

**Measures taken to deliberately mislead relevant decision-makers
about capabilities, intentions, or operations
in ways which may be exploited**

- **It is considered part of Information Operations (IO) and/or Command & Control Warfare (C2W)**

- **It is virtually always conducted *to further or cover another activity***

- **Historically, it has been used to great effect, and often is instrumental in achieving surprise**

What is deception?

The box depicts a doctrinal definition of deception taken from the appropriate literature on the subject: Joint Pub 3-58, Army FM 90-2, and Army FM 101-5-1. We've altered the definition slightly for breadth, replacing "enemy" with "relevant," in order to point out the potential utility of targeting noncombatants in addition to principals in the enemy command structure.

Notice what *isn't* present in the definition:

1. Offense or defense: Deception may be employed for both.

2. Friendly versus hostile: We should be prepared to counter an opponent's deceptions, *and* we may effectively employ deception ourselves.

3. Technology: Deception efforts may exploit technology, but they are not dependent upon it.

While deception operations and deceptive activities are actions in their own right, they are almost always conducted concomitantly with another activity. If history is any guide, deception can prove invaluable.

Military Deception — Process & Means

Deception Planning Process

Objective | Target | Story

Means of Deception

- Camouflage/ Concealment/ Cover
- Demonstration/ Feint/ Diversion
- Display/ Decoy/ Dummy
- Mimicry/ Spoofing
- Dazzling/ Sensory Saturation
- Disinformation/ Ruse
- Conditioning

How does deception work?

The planning process begins with the desired end and maps backward. We begin by asking, "What do we want to accomplish?" and then, "What do we need the target to do to gain that objective?"

Knowing what we want the enemy to do, we then ask, "Who can make that happen?" This is the *target* of the deception, and we subsequently ask, "What does the target need to believe in order to galvanize the desired actions?"

Once we know what the target needs to believe, we then ask, "What does the target need to *see* in order to engender those beliefs?" The answer is the *story* that must be told to the target, and it is told through the means of deception: the classic instruments such as camouflage or disinformation that comprise the deceiver's arsenal.

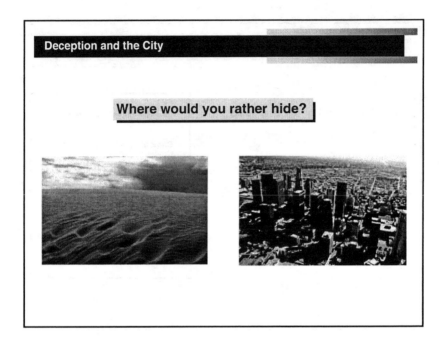

This may be glib, but it is a compelling point nonetheless: deception is different in different environments. It takes different forms, requires different efforts, and has different results. The RAND Arroyo Center Urban Operations Team has been considering the relationship between urban terrain and deception, and we have generated six hypotheses as part of our ongoing work on the subject.

RAND Hypotheses Concerning Deception & Urban Operations

How the City *Aids* and *Alters* Deception Efforts

- *Density*: Scope of Deception Is Increased

- *Activity*: 'Background Noise' Is Significant

- *Resources*: The Variety and Detail of Deceptions Are Increased

- *Op Tempo*: Decision-Making Is Hastier, Less-informed

- *Population*: Great Difficulty in IFF, Combatant Vs Noncombatant

- *Clutter*: The Edge of Intelligence-Gathering Is Blunted

Essentially, it is our contention that urban terrain both *abets* and *alters* deception in these six fundamental ways.

RAND Hypotheses Concerning Deception & Urban Operations

How the City *Aids* and *Alters* Deception Efforts

- *Density*: Scope of Deception Is Increased

- *Activity*: 'Background Noise' Is Significant

- *Resources*: The Variety and Detail of Deceptions Are Increased

- *Op Tempo*: Decision-Making Is Hastier, Less-informed

Given our time constraints, we present a brief discursion on four of them here.

Density: Scope of Deception Is Increased

Stalingrad, 1942

- Successful deception (*maskirovka*) covered Soviet offensive forces numbering 300,000 soldiers, 1,000 tanks, and 5,000 artillery pieces

- Germans (OKH) underestimated Soviet strength by 25–50% and were surprised

This is the canonical example of high-intensity urban operations, and it serves as an excellent example of how vast deception operations can get. Consider just how enormous the Soviet deception plan was to cover the attack against the German 6th Army. It was successful, as evidenced by OKH records, and gained surprise at all operational levels for the Soviet forces.

We believe that the geography—the urban terrain—was essential in allowing such a massive deception plan to succeed.

Activity: 'Background Noise' Is Significant

Lima, 1980–1989

- Shining Path *deliberately* operating at "sub-threshold" level

- Peruvian intelligence bogged down preparing urban battlefield— doubling and redoubling assets

---

At the other end of the urban operations spectrum, we see an interesting lesson in the urban campaign of the Peruvian "Shining Path" insurgency. Shining Path documents indicate that they deliberately sought to cloak their activities in the noisy environment of the city (e.g., was that a bank robbery or what the insurgents would call an "expropriation"?)

The effects of this strategy were that Peruvian intelligence was again and again forced to expand its assets in the city to try and isolate the "signal" from the "noise".

Resources: The Variety and Detail of Deceptions Are Increased

Belfast / Londonderry, 1969–1989

- Over time, Provos (PIRA) have employed every means of deception, often in combination (including nested deceptions), and often in an evolutionary "ruse race" with Brits/RUC

- Provo deceptions are both tactical *and* strategic

Consider an urban insurgency *campaign* as not a loosely linked series of terrorist acts but rather a form of military campaign evolving over time. One such source of material, which we can only scratch the surface of here, is the longstanding conflict in Northern Ireland.

During the urban phase of the IRA struggle against British forces in Northern Ireland, nearly every manner of deception imaginable was used at one point or another, tapping the wealth of resources available in the city. These resources include areas with a friendly non-combatant population!

It is also worth pointing out that deception in a terrorist campaign generally falls into operational levels like a military campaign. At the tactical level, deception aids the execution of terrorist acts (using disguises, diversionary explosions, etc.). At the strategic level, deception embraces those acts within a social/political/legal context (using hoaxes, false-flagging, disinformation, etc.).

Note that the use of deception by British (or Ulster Loyalist) forces is another important and fascinating topic, but due to time constraints is not treated here.

Op Tempo: Decision-Making Is Hastier, Less-informed

Algiers, 1954–1962

- FLN fighters use disguise and decoying actions to draw small units into ambushes and sniper fire

- Akin to PIRA tactics, the FLN desired to draw small French units into friendly ground (the Casbah) where information flowed in one direction only

---

Another illustrative example of deception employed in an urban insurgency campaign is the anticolonial struggle for Algerian independence.

The insurgent FLN force explicitly sought to isolate small French units, reasoning that those units would have less access to intelligence and thus be more gullible. They also sought to operate as rapidly as possible to encourage poor French decisionmaking by unit commanders, also increasing their gullibility.

It is important to point out that (reminiscent of the PIRA) FLN operational plans deliberately sought to operate on friendly ground whenever possible, knowing that the flow of HUMINT would be unidirectional.

RAND Ongoing Work

- *Density*: Scope of Deception Is Increased
- *Activity*: 'Background Noise' Is Significant
- *Population*: Great Difficulty in IFF, Combatant Vs Noncombatant
- *Resources*: The Variety and Detail of Deceptions Are Increased
- *Clutter*: The Edge of Intelligence-Gathering Is Blunted
- *Op Tempo*: Decision-Making Is Hastier, Less-informed

Deception Can Play A Significant Role in Force-Projection

One preliminary conclusion reached by the RAND Arroyo Center MOUT project is that these factors affecting and abetting the use of deception in urban operations *can apply to all contenders.* The classic model would have deception used by the force that knows and/or holds the city, but we believe otherwise.

Preliminary Conclusion

Deception Can Play a Significant Role in Force-Projection

US & Allies, 1999	Soviets in Berlin, 1945
• Operating almost exclusively abroad	• Soviets successfully use deception *in Germans' own backyard*
• Very much concerned with reducing casualties	• Soviets expressly use deception *to reduce casualties*
• Desiring tactical and operational surprise	• Soviets use deception to gain *surprise* at the operational and tactical levels

It is our contention that deception can be a very valuable commodity to a force-projection power that desires to keep casualties low in urban operations at a variety of intensities, and whose doctrine calls for tactical and operational surprise even when strategic surprise is forfeit.

We think that the example of the Soviet assault on Berlin in the closing days of World War II is an excellent example of leveraging deception to great effect for exactly those ends.

ANNEX 6: Q & A FROM INITIATIVES/TECHNOLOGY PANEL

Question: I have a question on C4I. Once a close battle in a MOUT environment has been joined, how will command and control at the brigade or the task force level provide a common operational picture? Are you doing anything in that arena?

Major Offen: What we have here is essentially a difference in echelon. The largest force we will look at is a battalion. That will occur both during the second joint experiment in September 1999 and the culminating demonstration in September 2000. We're not looking at digital systems and how they contribute to situational awareness. It's just outside the scope of what we're doing.

LtCol Schattle: We are. The command and control study includes scenarios with a brigade-sized task force, and we're looking at all of that.

Major Offen: A fair amount of that should come out of the JCFAWE, which is a joint AWE. It will include Marine and Air Force play as well.

LtCol Allison: During our AWE, we had the experimental combat operations center for our command element aboard *USS Coronado*. During an actual operation we might not put it on that platform, but it would be at sea. We had success passing operational status information from the *Coronado* to squad leaders through their user terminals. The information is shared but not necessarily common. At the highest level the battlespace is huge. The overall force commander was looking at ongoing actions from Concord Naval Weapon Station to Moffet Airfield. Then we broadened his battlespace using the joint conflict tactical simulations (JCATS) to give him more at the tactical level. Was it all relevant to him? Yes, it was. He responded with faster fires and better support. Do we need to establish more protocol with it? Absolutely. We're only skimming the surface in the experimentation process. The battalion commander in that battlespace also had his own combat operations center. We have tied those pieces together with the portion of the picture relevant to squad leaders' needs.

Question: Regarding deception, I was interested and surprised by your conclusions that seemed to indicate that deception could best be used by an expeditionary force. It seemed to me during the presentation that the urban environment is really set up for deception use by the defending adversary. Do you agree or disagree?

Mr. Gerwehr: I would actually say it could be used effectively in support of offensive and defensive operations. However, if you own a city, have knowledge of its terrain, and control its noncombatant population, you have significant advantages if you wish to employ deception. I strongly believe that there is room for the employment of deception during force projection. I tried to sample history in a way that covered the spectrum of intensities, the different kinds of missions, and also to demonstrate both friendly and adversary uses of it.

Question: I have a question on the MOUT ACTD process. It seems to me that you took a look at emerging technology in 1998, 1999, and a little bit in 2000. You had a frozen snapshot in time, and then you moved forward with selected items that you felt might be useful. Technology is evolving very quickly today. How do we avoid taking a snapshot of technology and saying "Well, that's it folks; we're going to go with it" when in fact it's evolving very quickly and we should be continuing to seize those golden nuggets on the path toward 2025?

Major Offen: I would say that the entire ACTD process is an effort to do that, to make an effort to shorten the acquisition timeline. Granted, we are sort of taking a snapshot. LtCol Allison, my Marine counterpart, described this is as the noncommissioned officer's ACTD. We have a fair number of very low-tech items that won't age as rapidly as some of the more high-tech systems. I think that regardless of the rate at which they age, there will be insights gained.

Question: I guess that was what I was trying to get at. It's not so much what might be available right now, but to avoid exactly that, to avoid having a guy run around in 2015 with what was great technology in 2000. We need to develop a process that keeps bringing this stuff to the surface to look at, not necessarily to buy.

Major Offen: I understand. I think the ACTD process does that quite well.

Question: Are you, in fact, open to introducing new technologies?

Major Offen: Absolutely. There are ACTD nominations every single year. This is an ongoing DoD process.

Question: Maybe just to rephrase the question for either of the first two briefers [Major Offen and LtCol Allison]: First, how do you see the relationship between the MOUT ACTD and the Urban Warrior AWE? Second, when you run across a technology that isn't quite ready for prime time, doesn't make the 70 percent cut, or whatever, what happens to it? Who else hears about it? Does someone else get a chance to look at it for later consideration?

Major Offen: Our Marine brethren have quite generously invited us to come and observe their Urban Warrior AWE and we've done that. Understand that Urban Warrior is a Marine effort, a Marine advance technology and digitization effort. The Army sort of has its own. However, I feel fairly certain that there's a lot of information sharing going on. I know that between the Dismounted Battlespace Battle Lab and the Marine Corps Warfighting Lab there's a fair amount of it. With respect to your follow-on question, the information and data that we collect, and the systems we look at, are readily available for review by a wide community. I think that community is fully aware of what we're doing. And as a matter of fact we have an upcoming event that's focused on looking at some things that are in the brass board stage. It is our advanced concepts excursion. We're inviting users from the Marine Corps, the United States Army, and the special operations community to come look at these systems. We are going to tell them not to consider only what they are seeing, but rather to use their imaginations and look at the future capabilities of the systems as well.

LtCol Allison: Sometimes people do get confused about what the Marine Corps MOUT ACTD program is doing, what we're doing in Urban Warrior, or what will happen during our next Capable Warrior series at the lab. The ACTD's purpose is to get things fielded rapidly down to the user level. We in the Marine Corps only have only one battle lab, and that's us, the Marine Corps Warfighting Lab. We also have a section that's called the Next Marine Corps that looks way out there, much like your Army After Next. We are always looking at how we tie the two together.

Audience comment: I'm a reserve officer and also a Chicago police-man. I hear a lot of great things going on here. I think we're missing out on maybe not putting enough emphasis on psychological operations and civil affairs. The Russian officer in my seminar [at the Army War College] said that Grozny was a psychological operations failure.

Audience comment: I really liked the piece on deception. I'm a military police officer so I understand where you're coming from. I like the part about the high-tech stuff because it helps with population and resource control. I think that if we get into a fight like you all are describing, and I feel certain we will, the enemy is going to hide, reduce our capability to get them, and is going to use civilians to strategic advantage. There's another thing. I think we have to be careful when we say we don't want to destroy infrastructure. I think that no matter what we do with respect to MOUT, we need to tell the enemy that we will look for him and we will kill him.

Question: Are you anticipating or encountering any friction in the joint arena as you seek to transition from experimental concepts to actual capabilities?

LtCol Allison: My sense is that if the world is changing and we're really serious about adapting to the world, we need to have a serious conversation with everybody in the joint community to make sure we know what it is we're really doing. The Marine Corps can clearly not go forward with a Marine-only system. That word should be passed. I think we need to find out what works and what doesn't work and then get word to the joint community. This is real easy to do with the MOUT ACTD because it's by nature a joint effort.

Question: How do you feel about psychological operations during MOUT? Are the Marines looking at developing their own capability to do that? Are you satisfied with the joint capabilities?

LtCol Allison: We're going to get a little money to fund nontraditional operational studies next year, and we want to concentrate on those skills that haven't been given a lot of emphasis, like PSYOP, civil affairs, public affairs in the operational sense, and humanitarian affairs. We're going to bring in individuals who have done this successfully, noted experts in the field. The battalion commander of our experimental force has some thoughts on what he calls vapor tactics. These involve the use of a combination of PSYOP and nonlethals to psychologically and physiologically turn up the heat on the bad guys so that by the time you try to collapse their defenses they are pretty well discombobulated.

RECENT MOUT OPERATIONS PANEL PRESENTATIONS

ANNEX 1: HAITI
COL David Patton, USA

OPERATION UPHOLD DEMOCRACY
Use of Military Police and Non-lethal Weapons in Haiti

Presented by
COL David L. Patton, U.S. Army
14 April 1999

My name is Dave Patton and I've been asked to discuss the role that military police soldiers and nonlethal weapons played during Operation UPHOLD DEMOCRACY in Haiti.

Operation UPHOLD DEMOCRACY
(Haiti)

U.S. Involvement

Phase I: Multinational Force Operations
(September 1994 – March 1995)

Phase II: UN Mission in Haiti (UNMIH)
(April 1995 – March 1996)

Phase III: U.S. Support Group Haiti
(April 1996 – Present)

I'll first address the limit of my personal experience in Haiti and its impact on my observations. I arrived in June 1995 at the request of the UN Force Commander, MG Joe Kinzer, to serve as the UN Executive for Haitian Security. In that role I provided both General Kinzer and the Senior Representative of the Secretary General (Mr. Lakhdar Brahimi, the senior UN civilian in country) overwatch of the programs under way to rebuild the Haitian criminal justice system: police, prisons, and courts. After four months in this position, General Kinzer asked me to stay and command the U.S. Support Group Haiti, which was to be the bilateral U.S. military contribution to the effort once the United States ceased its involvement in the UN Mission in Haiti. I took command of the Support Group in November 1995, led it through the departure of the U.S. UNMIH forces in March 1996, and relinquished command in October 1996. Thus I was not in Haiti during Phase I and, because of my position on the UNMIH staff, my observations of the military police during Phase II were from the vantage point of an interested observer rather than an active participant. The nonlethal weapons program began in Phase III.

Operation UPHOLD DEMOCRACY (Haiti)

Military Police Involvement

Phase II: MP Battalion in U.S. Contingent

This chart portrays the military police organization within UNMIH. The MP battalion was assigned to the U.S. contingent as part of the task force structure of both the 2nd Armored Cavalry Regiment and later, the 1st Brigade of the 101st Airborne Division. The MP company was a standard combat support unit with all of its firepower, mobility, and communications assets. The Indian MP Company was actually a reserve component crowd-control unit that was well trained and disciplined, but came with few vehicles or weapons and little communications equipment.

Operation UPHOLD DEMOCRACY (Haiti)

Military Police Missions

- **Patrolled critical areas of Port au Prince**
  - **Central City**
  - **Petionville**
  - **Carrefour**
- **Maintained U.S. presence in police stations**
- **Maintained U.S. presence in prisons**

Military police were used to patrol the most critical areas of the Port-au-Prince metropolitan area. The central city contained the principal shopping district for the majority of the residents. Petionville was a relatively wealthy enclave that was home to the Haitian elite, the diplomatic corps, and the upscale business establishments that still existed in the city. Carrefour was a portion of the city on the south side that harbored a significant segment of the disaffected population. MPs provided critical overwatch of police and prison facilities during the transition phase from the removal of military regime personnel, through the period in which these facilities were manned by interim elements, and into the introduction of newly vetted and trained professional police and prison guard forces.

Operation UPHOLD DEMOCRACY (Haiti)

Observations

- MP teams were properly configured for urban patrolling

- MPs commanded respect, but were able to build rapport

- MP presence deterred IPSF and HNP misconduct

- MP presence deterred prisoner abuse in prisons

U.S. MP units normally operate in three-soldier teams: NCO team leader, driver, and gunner. Use of two-team patrols became the norm and, because of their lack of organic mobility assets, Indian MPs were integrated with the U.S. MPs for patrol operations. Larger Indian MP elements functioned as a response force, prepared to move to incidents anywhere with the U.S. AO. Because of their police training and habits, MPs were able to quickly establish respect among the population in their role as peacekeepers while also developing those relationships that provided a continuous source of "street intelligence." In many situations, the mere presence of U.S. MPs in overwatch positions prevented Interim Public Security Forces and, later, the Haitian National Police from overreacting or losing control of situations in the stations and on the streets. Likewise, the continuous MP presence in the National Prison was credited by many international observers with preventing violence directed against prisoners by Haitian guards and with quelling several uprisings.

Operation UPHOLD DEMOCRACY (Haiti)

Non-lethal Weapons (NLW)

- **Began use with first Support Group Security Force (USMC FAST Platoon)**

- **Threat primarily criminal**

- **Munitions provided by Army Materiel Command**

- **Training provided by Army Military Police School MTTs**

Because the threat against the Support Group was postulated to be primarily of a criminal nature, the use of NLW was considered to be a viable option for both the defense of U.S. facilities and for crowd control. Each time the security force would rotate (generally every three to four months), a MTT would come down and take the unit through the training described in the next slide.

Operation UPHOLD DEMOCRACY (Haiti)

NLW Training Concept

- **Classroom instruction**

- **Range firing/familiarization**

- **Situational Training Exercises (STX)**

- **Reinforcement by unit leadership**

From a purely mechanical standpoint, the preliminary marksmanship instruction in the classroom and the actual shooting on the range gave the marines and soldiers a great deal of confidence in the NLW munitions. Perhaps the most important training, though, was the use of Situational Training Exercises (STX) crafted by the Support Group Provost Marshal and Staff Judge Advocate to test the judgment and temperament of both individual sentries, small units, and leaders.

This chart depicts the munitions that we had available to us in Haiti. In my 16 months in Haiti, we never used one round outside of training.

ANNEX 2: NEOs
Col T. W. Parker, USMC

NONCOMBATANT EVACUATION OPERATIONS

"SHARP EDGE"

COLONEL T.W. PARKER, USMC

OPERATION SHARP EDGE

- CLASSIC MEU (SOC) OPERATION
- PROTECTION AND EVACUATION OF NONCOMBATANTS
- SECURITY OPERATIONS
- SHOW OF FORCE

I'm not going to bog down on details regarding Liberia and Operation Sharp Edge today. I want to say that I think Liberia is the kind of a case study that represents a classic NEO operation; it took place in a city, in an American embassy, and in an urban area accessible by ships. It had all of the pieces that we read about in the doctrinal publications. There's only one problem: doctrine must remain flexible because it begins to fall apart during execution.

MISSION

- SECURE THE US EMBASSY
- EVACUATE US NATIONALS AND DESIGNATED FOREIGN NATIONALS
- PROVIDE LOGISTICS SUPPORT TO THE EMBASSY

REQUIREMENTS

- FWD CMD ELEMENT ASHORE
- INFORMATION
- MAPS
- PHOTOS
- 3D PICTURE OF AO

I've kept this fairly simple. I want to deal with impressions and bigger-picture items rather than the details of this operation. One of the things that was absolutely critical to our success in the Liberia NEO was having a forward command element ashore. Somebody has to be there on the ground to be able to talk directly back to the command. We were lucky enough to be able to get in one of the last fixed-wing aircraft that entered Liberia. Were able to get our unexploded ordnance (UXO) and his forward command element ashore in the embassy. Now we had practiced and rehearsed with twenty people or so. The actual forward command element we put ashore was six individuals. One of these six was especially critical: a captain who was a CH-46 helicopter pilot. He became absolutely essential during the operation because he could tell us what LZs looked like, what kind of helicopters could go where, and what fit given LZs. There are a lot of plans out there. There are a lot of people out there who can tell you that you can land on the tennis court, you can land on the basketball court, you can land in between these buildings; we've measured them and we think they're good. Our helicopter pilot was able to tell us exactly what it looked like, and, by the way, the embassy plans were wrong. Where they said you could land, you couldn't. Places where they said you couldn't land, you could. New information is a critical requirement. However, there is such a thing as too much information. We were finally at that point. We got flooded. Our little S2 shop simply got flooded with information. The problem became sorting it out and trying to figure out which pieces of it were really important and which were irrelevant. Maps—we got flooded with maps. The lesson learned from Grenada was maps. We had 2,200 people; I think everybody had a map by the time it was over with, so maps did not become a problem. What was also critical to us was the ability to have a three-dimensional picture of Monrovia and of the embassy in particular. We were able to get them by employing two methodologies. One was taking photos and maps and having our topographic platoon detachment on board ship construct scale models of the city and the embassy. Secondly, computer-generated video was used to show us routes into the city, particularly air routes. Putting all of that together, we felt as at home in Monrovia as we would have walking out the back door. That is absolutely true.

PLANNING

- COMMUNICATIONS
- INTELLIGENCE/THREAT
- INFANTRY
- MEDICAL
- HELICOPTER AVAIL
- SCREENING

An earlier presentation addressed communications. I think the key here is to have a backup to the backup and also to design a plan that can be executed without communications. That means it has to be simple and everybody has to know what it is. There have also been discussions here about infantry requirements. The requirement for infantry is probably larger than what is on hand. We immediately turned our artillery battery into a provisional rifle company since we weren't going to use artillery. Essentially, you can't have too much infantry. A very small embassy like the one in Monrovia is probably not much larger than this building. We had a battalion of four rifle companies plus a provisional rifle company. The requirements for infantry were extensive.

When you are dealing with hundreds and thousands of civilians, you have to have the ability to treat their medical problems. We did not have enough doctors. The ability to treat mass casualties is important. Medical personnel need to have rehearsed what they are going to do in those mass casualty situations.

Helicopter availability was also a consideration in our planning. We always planned to have fewer helicopters available than we actually needed. That way when they broke we had a plan prepared.

Screening of evacuees—there is much made of screening evacuees in the doctrinal publications. I've got to tell you that screening evacuees on the ground and trying to sort out who's a good guy and who's a bad guy in a group of five hundred to a thousand individuals doesn't work. You have to do a very cursory screening. Put them on helicopters, get them out of the area, get them to a safe place, and then screen. The ones you don't want you can take back. We found one guy who wanted us to take him back. He was an Iraqi. In August 1990, Iraqis were no longer people that we wanted to evacuate.

LESSONS LEARNED

- INFANTRY REQUIREMENTS
- THERE IS NO PERMISSIVE ENVIRONMENT
- SPEED
- SIMPLE PLAN THOROUGHLY REHEARSED
- FLEXIBILITY

When dealing with cities with their many small areas, buildings, and other places where people can hide, your infantry requirements go up. When you need to handle hundreds of evacuees, your infantry requirements go up. Regarding the second bullet there, there is no permissive environment. I know that even in the Marine Corps, at that point anyway, we were talking about permissive NEOs and non-permissive NEOs. We went by the rule that there is no permissive environment. There is no such thing. Plan for it not to be permissive. If it is, that's great.

Speed is critical during a NEO. There are people shooting in all directions at anything that moves, and the more rapidly you get your people out, the more rapidly you move them back to a safe area and the better off you are. As I said, screen those people on board ship; screen them in a safe area. Do it somewhere other than in the landing zone.

A simple plan thoroughly rehearsed and flexibility are key. We developed very simple plans that were rehearsed over and over and over at sea. We rehearsed them over and over and they were flexible. The Sharp Edge plan for Liberia, we had a half-dozen separate subsets of evacuation plans. None went according to the plans that we had when we left the ship, not one of them. But all of them were such that we were able to adapt the plan that we had rehearsed.

The last point that I would make, and I don't have it up here, regards people. As we talk about technological breakthroughs, the ability to see through walls and nonlethal weapons and communications gear and all the rest, I think that the most impressive thing about our experience with marines who came under fire for the first time was how well they did. General Krulak talks about his strategic corporal and what the requirement is for the strategic corporal: a guy who's smart, physically fit, and one who will make decisions in difficult environments that could have strategic implications. The strategic lieutenant is also part of that. We have to be mindful that there will be NCOs and lieutenants who will make strategic decisions in front of CNN that will affect National Command Authorities decisions, decisions that will affect the way forces are used. I think that when we talk about MOUT operations and NEOs the importance of people is absolutely critical: having the right people, building the right force with the right people. Thank you.

ANNEX 3A: GROZNY I: ASYMMETRY IN URBAN WARFARE
Mr. Timothy Lee Thomas, FMSO

Asymmetries in MOUT:

The Case of Grozny

Asymmetry

(1) Not symmetrical;

(2) the unexpected, non-traditional approach

(3) using offsets against a stronger foe (terrorism; PSYOP; WMD; computer attacks; guerillas, etc.)

Former Chechen Vice President Yanderbeiyev on the Battle of Grozny:

"the situation did the organizing"

Urban Warrior 1997

"the difficulty of the fight would increase as the organization of the opposing force decreased"

Urban Combat: Then versus Now

• Diplomacy follows	• Diplomacy leads
• sledgehammer	• vapor tactics
• artillery preparation	• artillery support
• tanks in the lead	• tanks in support
• surface discipline and training	• psychological discipline and training
• linear attrition	• avoid linear attrition
• no maps	• pagers
• infantry walk	• infantry ride on discs

The Nature of Urban Combat and the Future of US Forces

Urban Combat Requirements	Force 21 Characteristics
• Large force	• Small force
• slow pace and tempo	• dominant maneuver
• long duration	• precision fire
• long prep fires	• logistics
• constant comms	
• LOS constraints	

ANNEX 3B: THE BATTLE OF GROZNY: DEADLY CLASSROOM FOR URBAN COMBAT[1]
Mr. Timothy Lee Thomas, FMSO

> Best policy in war—thwart the enemy's strategy;
> second best—disrupt his alliances through diplomacy;
> third best—attack his army in the field;
> worst strategy—attack walled cities.

—Sun Tsu, *The Art of War*

The battle for Grozny, the capital of the small Russian Republic of Chechnya, took place in January 1995. It pitted a hastily assembled and unprepared Russian force against a Chechen force of regulars and guerrillas equipped with Russian weapons and a belief in their cause. The Chechens held their own for three weeks but eventually lost the city to the Russian armed forces in late January (the Chechens retook the city in August 1996).

Both sides learned or relearned many lessons of urban combat, most of them the hard way.[2] This article examines the most important of those lessons, the interesting and perhaps surprising conclusions drawn by the Russians about modern urban warfare, and their implications for U.S. soldiers and urban warfare theory.

Background

The Russian Republic of Chechnya is located in the southeastern part of Russia near the northwestern end of the Caspian Sea. Chechnya declared its independence from the Soviet Union in October 1991.

---

[1]This material appeared as Tim Thomas, "The Battle of Grozny: Deadly Classroom for Urban Combat," *Parameters*, Vol. 29 (Summer 1999), pp. 87–102 and is printed with the kind permission of the editors.

[2]For an extended, blow-by-blow version of the fighting during the month of January, see either Carlotta Gall and Thomas de Waal's firsthand accounts in *Chechnya: Calamity in the Caucasus* (New York: New York University Press, 1998), chapters 1 and 10, or Timothy Thomas, "The Battle for Grozny," *Slavic Military Studies*, Vol. 10 (March 1997), pp. 50–108.

This declaration by Chechen President Jokar Dudayev was not unexpected; the region's history is scored by episodes of intense Chechen-Russian battles that encouraged hatred toward Russia and a desire for independence. Further, Russia was in disarray at the time, with then Russian Republic President Boris Yeltsin in confrontation with Soviet President Mikhail Gorbachev over the issue of sovereignty. Yeltsin encouraged Soviet republics (but not semi-autonomous ones like Chechnya) to "take all the sovereignty they could swallow." Dudayev interpreted Yeltsin's words to fit his situation.

Grozny had nearly 490,000 residents in 1994. It included many multiple-story buildings and industrial installations and covered some 100 square miles. (By comparison, the Joint Readiness Training Center for urban combat in the United States covers less than a tenth of a square kilometer, offering but one indication of how urban training can differ from reality.) A Chechen opposition movement developed in 1993, finally attempting to overthrow President Dudayev in late November 1994 through an armed attack. The attack was repulsed by Dudayev's forces. Russian complicity was at first denied by Moscow, but then acknowledged when Dudayev paraded several captured Russian soldiers before TV cameras.

The indignity and embarrassment over the exposure of Russian involvement caused Russian Federation President Boris Yeltsin, in his third year in office after ousting Gorbachev, to order troops to start moving into Chechnya on 11 December. Planners had less than two weeks to move and position forces and supplies. By New Year's Eve, Russian forces had Grozny surrounded on three sides and entered the city from the north, moving headlong into hell.

The first unit to penetrate to the city center was the lst battalion of the 131st "Maikop" Brigade, the latter composed of some 1,000 soldiers. By 3 January 1995, the brigade had lost nearly 800 men, 20 of 26 tanks, and 102 of 120 armored vehicles. For the next 20 days and nights, Russian artillery rounds rained down on the city, sometimes at the rate of 4,000 an hour.[3] Local residents left the city or took refuge in basements while the Russian armed forces fought Chechen

---

[3]David Remnick, "In Stalin's Wake," *The New Yorker*, July 24, 1995, p. 48.

"freedom fighters" or "bandits" (depending on one's perspective) on the streets and in the buildings above them.

When more captured Russian soldiers were shown on TV, the mothers of some went to Grozny to negotiate their sons' release. Those negotiations took place in the center of the city without Russian government assistance and while under Russian artillery bombardment. Dudayev extracted a promise from the Russian soldiers he released of eternal indebtedness to their brave mothers.

The struggle continued until 20 January, when the Russians finally took the city center and raised the Russian flag over the Presidential Palace.

Before delving into the lessons learned from this battle, we may note several contextual factors that conditioned the outcome.

The Chechen armed force spoke Russian, had served in the Russian armed forces, and had Russian uniforms. This made it much easier to understand Russian tactics and plans, and to use deception techniques. The Chechen force was not a typical army but rather a composite force of armed home guards (guerrillas) and a few regular forces. Much of the equipment in their possession had been left by Russia's armed forces in 1993 when departing Chechnya. By one account the Chechens had 40 to 50 T-62 and T-72 tanks, 620–650 grenade launchers, 20–25 "Grad" multiple rocket launchers, 30–35 armored personnel carriers and scout vehicles, 30 122mm howitzers, 40–50 BMP infantry fighting vehicles,[4] some 200,000 hand grenades, and an assortment of various types of ammunition.

The Russian armed forces that attacked Grozny, while well equipped, were not the same professional force that opposed the West during the Cold War. Russian Minister of Defense Pavel Grachev, in a top-secret directive, listed some of the problems of his armed forces just ten days before the start of the war. He noted that the combat capabilities of the armed forces were low, the level of mobilization readiness was poor, and the operational planning capability was inadequate. Soldiers were poorly trained. Their suicide rates as well as

---

[4]Sergey Surozhtsev, "Legendary Army in Grozny," *Novoye Vremya*, No. 2–3 (January 1995), pp. 14–15.

the overall number of crimes in the force were up. Knowing the situation so clearly, Grachev's bold prediction that he could take Grozny with a single airborne regiment in two hours is incomprehensible.[5] Perhaps Grachev privately understood the true problems in the force but put on the face of public bravado to support the presidential directive he had received.

Other analysts confirmed the dismal state of readiness, estimating the capability of the Russian armed force to carry out combat missions as five or six times lower than what it had been in just 1991.[6] Not only was the force poorly trained, it also was undermanned. On the eve of the operation, Grachev apparently had a force of some 38,000 men, only 6,000 of whom entered Grozny on New Year's Eve. Dudayev is believed to have had 15,000 men in Grozny. This means that the 6:1 force ratio desired for attacking a city (a doctrinal norm derived from combat experience in World War II) clearly was not attained. On the contrary, the correlation of forces was 1:2.5 *against* Russian forces at the start of combat. In addition, the force that entered Grozny was a composite force, with some battalions composed of members from five to seven different units. Crews often hardly knew one another. One Russian officer noted that a rehearsal for taking a built-up area had not been conducted in the last 20 to 25 years, which contributed to decisions such as sending the force into the city in a column instead of in combat formation.[7] These facts, combined with the bad weather, the hasty political decision to enter the city, and the lack of training, offered the Russian force little chance for quick success.

Lesson One: Know Your Opponent and His Turf

Societies are run by different methods. Some are governed by the rule of law, others by the rule of men. Some are governed by reli-

---

[5]N. N. Novichkov et al., *Rossiyskiye vooruzhennyye sily v Chechenskom Konflikte: Analiz, itogi, vyvody (analiticheskiy obzor)* (Paris, Moscow: Holveg-Infoglob-Trivola, 1995), pp. 18, 19.

[6]Ibid., p. 21.

[7]Vitaliy Kudashov and Yuriy Malashenko, "Communications in a City," *Armeyskiy Sbornik*, No. 1 (January 1996), p. 30.

gious or local tradition, and still others by the tradition or customs of the clan. Chechnya was a society run by the rule of the clan.

Two traditions of the clan that unify the Chechen people are *adat* and *teip*. *Adat* is an ancient system of retribution, an unwritten code that is followed more closely than the Russian penal code or other imposed civil laws. The code is reputedly based on revenge, incorporating an "eye-for-an-eye" sense of justice. For example, after two of their comrades had been killed, Chechen fighters took a building in Grozny and seized some Russian prisoners. They killed two and let the rest go.[8] They had their revenge.

The tradition of clan or tribe *(teip)* relationships is equally important and should have been stressed to Russian forces. *Teip* members fight fiercely to preserve their clan's independence, culture, and separate identity. Relations between *teips* "are based on blood feuds."[9] There are more than 150 *teips* in Chechnya, whose membership "ties a Chechen to a large extended family and to an ancestral piece of land."[10] If an opponent of the Chechens fails to take into account both *teip* and *adat* (as well as the long Chechen tradition of looking to older men for wisdom and to younger men for the "warrior" spirit), then he will not understand the fundamental issues uniting Chechen society and their will to fight foreign domination. Such will can outlast outside weaponry and presence. This lesson becomes more and more important to U.S. planners as the American armed forces move to an expeditionary posture. An outside force can't stay forever, and the will of the local populace may win in the end.

In Chechnya there was even more at work than these internal cultural factors, however. There also existed an intense historical hatred of Russia and Russians among elements of the population, a reaction to the lack of respect shown by Soviet leaders and their Russian predecessors. In 1816, for example, Russian Caucasus commander General Alexi Yermolov insisted that "the terror of my name should guard our frontiers more potently than chains or fortresses." He launched

---

[8]Peter Ford, "Chechens' Eye for Eye Vendettas Shape War," *The Christian Science Monitor,* March 8, 1995, pp. 1, 13.

[9]Novichkov et al., p. 5.

[10]Gall and de Waal, p. 26.

a scorched earth policy, treating the Chechens with extreme cruelty to perpetuate his claim. In 1949 Soviet authorities erected a statue of General Yermolov in Grozny. The inscription read, "There is no people under the sun more vile and deceitful than this one."[11] This unbelievably callous and calculated insult by Soviet authorities ensured the eternal hatred of many Chechens, demonstrating how poorly Soviet authorities understood their own people. During the 1970s and 1980s, the Chechens repeatedly attacked this statue.

Joseph Stalin earned the further enmity of the Chechen people by deporting the entire population to Central Asia in 1944. Many died during these deportations, which Chechens viewed as genocide. They returned to their homeland 13 years later during the premiership of Nikita Khrushchev.

Yeltsin and his military planners failed to consider the "receptivity" of the people in Grozny to their demands and intentions. The Russian army lacks civil affairs units, and this missing element compounded its problems. In the same manner, any force considering an attack in an urban environment must evaluate both the type of opponent it is attacking (guerrillas, regular forces, etc.) and its will. If the opposing force has deep and persistent antipathy toward the attackers, then it will be impossible to achieve victory without a decisive confrontation and military conquest. The local force has the advantage; if it can persevere, it can pick the attacker apart in both the short and long term, eventually wearing him out. In this sense, the moral-psychological orientation of the defenders adds an important element beyond mere weaponry to the "correlation of forces."

In addition to understanding one's opponent, an attacker must know the urban terrain over which he will fight. The Chechens obviously had a huge advantage in Grozny, as does any native defender. Not only did they know the city's sewer, metro, and tram systems intimately, they also knew the back alleys, buildings, and streets. Russian forces were not so prepared. They had 1:100,000 scale maps when a scale of 1:25,000 or even larger was needed. As a result they often got lost, finding themselves in Chechen ambushes or exchanging fire with friendly units. Chechens took down street signs and

---

[11]Ibid., pp. 39, 41, 62.

repositioned them in cleverly misleading positions to confound the Russians. Unit boundaries were almost impossible to coordinate because of the lack of adequate maps. Tactical maps were often made from plain blank paper by hand, with Russian soldiers filling in the sheet with the city vista (streets, buildings, etc.) in front of them.

Modern urban sprawl continues to make this aspect of military operations in urban terrain (MOUT) more appealing to the defender. The Chechens reverted to a battle of "successive cities" after the Grozny battle ended, hoping to recreate their Grozny successes elsewhere. They moved their operations base to Shali, Argun, and other city centers. They recognized that they could accomplish two things with this tactic: they could negate Russian advantages of firepower in the open from helicopters, fixed-wing aircraft, and tanks, and they could blend in with the local population to their advantage. This not only continued to make it difficult to distinguish combatants from civilians, but it also helped the Chechens get the local population on their side. This was usually the result when Russian forces entered a city, destroyed property and buildings, and killed or wounded civilians while searching for their armed opponent.

The average Russian soldier possessed neither the cultural savvy nor the street smarts for such confrontations. Since urban combat is resolved at the squad and platoon level, well-trained soldiers are essential. Too often this fact is ignored by forces contemplating an urban action. For Western armies, particularly expeditionary forces, there must be civil and public affairs units attached to help the soldier prepare for urban combat. A cultural understanding of the battlefield can greatly assist both the commander in understanding his mission and the average soldier in fulfilling it. An understanding of the city infrastructure offers similar advantages.

Lesson Two: Don't Assume—Prepare, Prepare, Prepare

When planning for the intervention into Chechnya, the Russian force made several questionable assumptions. First was that the Chechen force would not resist or stand up psychologically to the concentration of large groups of state troops. Soviet forces had succeeded in Czechoslovakia in 1968 with such an operation, and they may have banked on a repeat performance in Chechnya. A second assumption was that qualified planners were still present on the General Staff.

Much of the intellectual strength of the General Staff probably had atrophied along with the general dissipation of the armed forces over the previous five years. But even if the General Staff was fully operational and capable, any operation of this complexity may have been beyond its ability to prepare in two weeks (in preparation for the Czech invasion, for example, there were at least six extensive Warsaw Pact exercises over the course of several months to practice ground operations, air defense, logistics, and other elements before the intervention).[12] Finally, the plan presupposed a trained, coherent force that was as capable as the old Soviet military. None of the Russians' assumptions stood the test of reality.

Many outright errors were committed during the hasty preparation of the force as well. For example, the operations plan omitted technical support resources (such as communication equipment), and there was no coordinating agency linked with the president's administration to resolve political problems. The administration's information/propaganda machinery also failed to prepare the mass media to report positively on the reasons for the intervention or to illuminate the national interests at stake. Thus Russia lost the political and information battles in the first days of the conflict. Many of these problems were aggravated by the fact that at the time of the intervention, Russia did not have a national security concept, and only an outdated military doctrine.

In addition, three powerful ministers (Defense, Internal Affairs, and Internal Security) all had troops in the fight but failed to integrate their efforts.[13] As one source noted, "The enormous losses of the early days were caused by the poor level of professionalism of the command/staff element, which underestimated the enemy and was staggeringly negligent in coordinating actions among individual units and subunits as well as among the various types of forces."[14]

Other Russian problems included complacency as to the location of the main and reserve force, poor highway traffic control, a lack of

---

[12]Graham Turbiville, "Soviet Bloc Maneuvers," *Military Review*, Vol. 58 (August 1978), pp. 19–35.

[13]Novichkov et al., pp. 28–30.

[14]Ibid., pp. 30–31.

knowledge of the area, and no clear-cut troop instructions on how or when to use their weapons. Soldiers were sometimes prohibited from massing fires and lacked clear rules of engagement and target adjustment criteria. Some troops had just arrived from training units and had no idea how to operate as part of a unit. All they knew were individual soldier skills.[15] According to one report:

> In the 81st Motorized Regiment of the 90th Tank Division, out of 56 platoon commanders, 49 were yesterday's [civilian college] students. More than 50 percent of the men sent to war had never fired live shells with their tank cannons, and had no idea of how to do so. Military cooks, signalers, and mechanics were appointed to shoot antitank guns and missiles as well as machine guns.[16]

Just days before kicking off the operation into Grozny, a unit that was deployed at the Mozdok staging area conducted the following training: assembly and disassembly of equipment; range firing and field training; company tactical exercises and driving combat vehicles; battalion field training; driver testing; and alert drills.[17] Not a word about training on combat in cities.

Perhaps the most serious deficiency in the preparation phase was in intelligence data. The Russians had almost no information about the situation in the city, especially from human intelligence sources. Military intelligence did not delineate targets for air and artillery forces, and electronic warfare resources were not used to cut off President Dudayev's communications. Reconnaissance was poorly conducted, and Chechen strong points were not uncovered. There was little effective preliminary reconnaissance of march routes, reconnaissance amounted to passive observation, and reconnaissance elements appeared poorly trained.[18] Simply put, the Russians did not do a proper intelligence preparation of the battlefield—indeed, there does not seem to be an established procedure for processing data for the intelligence preparation of the battlefield in the Russian armed forces. Commanders and troops tried to overcome this short-

---

[15]Ibid., pp. 23, 25.

[16]"Soldiers Sent to Battle after 8 Days' Training," *Moscow News*, April 17–23, 1997, p. 5.

[17]Ibid.

[18]Novichkov et al., pp. 34, 35.

coming in the course of combat actions, leading to delays in operations and reduced effectiveness.

These first two lessons (know your opponent and battleground, and the importance of preparation) may seem elementary, but they may also reveal aspects that U.S. planners are most likely to miss. Analysts writing urban doctrine should raise their focus from tactics to consider also overarching concepts such as political considerations, limitations of city fighting, worldwide integration of economic assets, characteristics and types of opposing forces (guerrillas, regular force, willingness to violate international law), city size and infrastructure, and probable enemy methods for negating U.S. operating superiority. It is a combination of these latter facts that will drive the tactics and operations when going into a city. Neither strategy nor tactics can be developed in isolation from them. U.S. forces thus need an urban combat courses-of-action methodology to help select the optimum approach to each situation.

Lesson Three: Choose the Right Weapons

The Chechen weapon of choice was the rocket-propelled grenade launcher (RPG). The RPG was most feared by the Russians because of its multiplicity of uses. It could be used to shoot over buildings like a high-trajectory mortar, and it could be used either as an area weapon when fired over troop formations or as a precision weapon when fired directly at armored vehicles. Some destroyed Russian tanks were hit more than 20 times by RPGs.

A second weapon of choice for the Chechens was not really a weapon at all. It was the multitude of information-technology gadgets, especially cellular phones and commercial scanner systems, that allowed the Chechens to communicate easily with one another, ensured the coordination of combat operations, and allowed Chechens to listen in on Russian conversations (thereby proving to be a force-coordination multiplier). On many occasions, the Russians felt the Chechens knew what they were going to do ahead of time, and for this reason believed these communication devices were like weapons. The Chechens also used mobile TV stations to override Russian TV transmissions and to deliver messages from President Dudayev directly to the people. The Internet was also used, especially to raise funds and assistance from abroad.

Flame-throwers appear to have been a weapon of choice for the Russian force. One article written after the fighting noted that the Kalashnikov assault rifle, the Mukha grenade launcher, and the Shmel flame-thrower were a "soldier's best weapons."[19] The flame-thrower was chosen as much for its psychological effect as its ability to flush people or snipers out of buildings at a considerable range. Evidence supporting the view that this is an important Russian weapon was provided when an improved, jet-powered model was advertised for sale abroad in October 1998. It reportedly was capable of the same effectiveness as 152mm artillery rounds, and had a maximum range of fire of 1,000 meters (over a half mile!).[20] With its portability and range, it may prove to be an adequate substitute where the use of supporting artillery would be difficult.

A "weapon" of choice for both Russians and Chechens was the sniper, who caused panic and havoc with just a few well-placed shots. There are reports that the Chechens employed female snipers from the Baltic region. Snipers were extremely effective in slowing a convoy's movement and forcing a column to take another route. One observer wrote:

> One experienced sniper is capable of doing what will prove to be beyond the capability of a tank, gun, or entire infantry subunit: disable a commander, destroy a gun or mortar crew, control one or two streets ... and, most important, instill in the enemy a feeling of constant danger, nervousness, and expectation of a sudden shot. Everyone fears the Chechen snipers in Grozny.... There are many cases where a sniper wounds a serviceman, and then kills the wounded person and those who come to his aid.[21]

The sniper could also use an RPG in conjunction with a sniper rifle. A real problem for Russian troops was identifying snipers who shot at

---

[19]Vladimir Berezko, "Flame Throwers: A Second Birth," *Krasnaya Zvezda*, December 29, 1995, p. 2. The Mukha antitank weapon was second in popularity to the RPG. There were also reports that the flame-thrower was underutilized, but these were deemed a mistake.

[20]ITAR-TASS, 0943 GMT, October 2, 1998, as translated by FBIS and downloaded to its Web page. The high accuracy is due to an engineering development in which the motor and fire satchel inside the bore (container) separate.

[21]Oleg Mikhaylov, "A Rare and ... Unpopular Specialty: About One Lesson of the New War," *Armeyskiy Sbornik*, No. 3 (March 1995), pp. 38–41.

them and then donned a Red Cross armband and mingled with the local populace and the Russian soldiers he was killing. To counteract this, Russian checkpoints began forcing the Chechen men to take off their shirts. Soldiers would look for bruises on the shoulder from weapon recoil, for powder burns on forearms, or for a silver lining around cuffs (from mortar or artillery propellant bags). They also smelled clothing for gunpowder and looked for traces of it under fingernails or on arms or legs. Russian forces also employed snipers, but not with the same degree of success as the Chechens. A March 1995 article decrying the neglect of sniper training attests to this fact.[22]

The correct mix and employment of weapons in the city were also important. Grozny was a three-tiered fight (upper floors of buildings, street level, and subterranean or basement), and the weapons had to fit. Russian tanks could not lower their main gun tubes and coaxial machine guns low enough to shoot into basements harboring Chechen fighters. To correct this problem, the Russians put ZSU-23-4 self-propelled, multi-barreled, anti-aircraft machine guns forward with columns to fire at heights and into basements.

The use of artillery and air power in the city was counterproductive in many instances. Indiscriminate bombing and shelling turned the local population against the Russians. The locals included some Russian citizens who were inhabitants of Grozny (and who found it incomprehensible that their own leaders had such disregard for the lives of civilians). Most of the Russian population of Grozny lived in the center of the city. Since this is where the most severe fighting took place, Russian civilian casualties were high.

Lesson Four: Adapt Tactics to the Situation

The principal Chechen city defense was the "defenseless defense." They decided that it was better not to have strong points, but to re-main totally mobile and hard to find.[23] (Some strong points did exist

---

[22]Ibid.

[23]Chechen Brigadier General Zelimkhan Yandarbiyev recently emphasized this point, noting that in the early fight for the city, "the situation did the organizing." "Chechen

but were limited to dug-in tanks, artillery, or BMPs to engage targets head-on.) Hit-and-run tactics made it difficult for the Russian force to locate pockets of resistance and impossible to bring their overwhelming firepower to bear against an enemy force. Russian firepower was diluted as a result and could be used only piecemeal. Chechen mobile detachments composed of one to several vehicles (usually civilian cars or jeeps) transported supplies, weapons, and personnel easily throughout the city. Chechens deployed in the vicinity of a school or hospital, fired a few rounds, and quickly left. The Russians would respond by shelling the school or hospital, but usually after the Chechens had gone. Civilians consequently viewed this action as Russians needlessly destroying vital facilities and endangering their lives, not realizing who had initiated the incident. The Chechen mobility and intimate knowledge of the city exponentially increased the effect of their "defenseless defense."

The slaughter of the Russian 131st Brigade was a result of this tactic. Russian forces initially met no resistance when they entered the city at noon on 31 December. They drove their vehicles straight to the city center, dismounted, and took up positions inside the train station. Other elements remained parked along a side street as a reserve force. Then the Chechens went to work. The Russian lead and rear vehicles on the side streets were destroyed. The unit was effectively trapped. The tanks couldn't lower their gun tubes far enough to shoot into basements or high enough to reach the tops of buildings, and the Chechens systematically destroyed the column from above and below with RPGs and grenades. At the train station, Chechens from other parts of the city converged on the station and surrounded it. The commander of the Russian unit waited until 2 January for reinforcements, but they never arrived. His unit was decimated.

The most lethal Chechen force in those early days of January was led by one of President Dudayev's most trusted warriors, Shamil Basayev. Basayev's "national guard" force consisted of some 500 men who had fought in Abkhazia against Georgians in 1992–1993. Battle-hardened, they moved in groups as large as 200 at times, showing up in cars with guns blazing.[24] The more typical Chechen

---

Commander on Modem Separatism," *Nezavisimoye Voyennoye Obozreniye*, No. 2, January 22–28, 1999, p. 2.

[24]Gall and de Waal, p. 205.

combat group was a three- or four-man cell. Five of these cells were usually linked into a 15- to 20-man unit that fought together.

Some Chechen soldiers pretended to be simple inhabitants of Grozny, volunteering to act as guides since it was so difficult to navigate in the City.[25] They subsequently led Russian convoys into ambushes. Russian forces tried to counter Chechen ambush tactics by using a technique called "baiting," in which they would send out contact teams to find Chechen ambushes. In turn, the Chechens used a technique called "hugging," getting very close to Russian forces. This technique eliminated the Russian use of artillery in many cases, and it exposed baiting tactics.

The Chechens were proficient at booby-trapping doorways, breakthrough areas, entrances to metros and sewers, discarded equipment, and the bodies of dead soldiers. Some command-detonated mines were also used, but this weapon found greater use in other cities the Chechens defended. (A detailed 1998 Russian article about the importance of initially using plenty of expert engineer-reconnaissance forces in MOUT was published to teach how to counteract such threats.)[26]

Russian forces became wary of moving into a building and learned to proceed methodically. They began taking one building at a time, and moving block by block instead of rapidly moving into the city center as they had at the beginning of the intervention.

Another significant Russian problem was the delineation of boundaries between units owing to the nonlinear nature of urban combat. For the Russian force, this problem was complicated by four factors: poor communications that prevented units from knowing where other units were; the absence of an integrated communications system tying together different units from the Ministry of Internal Security, the army, and the other services; different operational tempos in

---

[25]In addition, Chechen artillery observers operated in the rear of Russian forces disguised as peaceful residents or refugees. Some Chechens, especially Russian-speaking women, reported on Russian forces. Chechens also took hostages to coerce family members to serve as artillery spotters in the rear of Russian forces. *See Armeyskiy Sbornik*, No. 1 (January 1996), pp. 37–42.

[26]Vadim Mayatskii, "When Storming a City . . . " *Armeyskiy Sbornik*, No. 2 (1998), pp. 6268.

different parts of the city that caused one unit to get ahead of another; and dealing simultaneously with both vertical and horizontal boundaries within a building. This difficulty in ascertaining boundaries resulted in several incidents of fratricide and instances in which units were pinned down by friendly fire for up to an hour. Aware of these problems, the Chechens exploited boundary conditions whenever possible. To help overcome such difficulties, a Russian expert recommended that units wear pagers and use a map display system known as Cospas-Sarsat during future operations. (Cospas-Sarsat is a system of geostationary satellites that act as a global positioning system, especially for search and rescue.)

A final tactical issue was the Russian use of assault detachments and tanks to seize buildings and drive the Chechens from the city.[27] Initially the Russians relied heavily on tanks in Grozny, but this approach was soon abandoned, with infantry and marines then becoming paramount. The initial instruction pamphlet issued to Russian soldiers in Grozny noted that a tank platoon should move at the head of the column, covered by motorized riflemen and flame-throwers. Reserve teams advancing in armored personnel carriers behind the tanks would fire against second and third floors. Three months later conflicting advice appeared in Russian army magazines. Tanks were advised to seal off city blocks, repel counterattacks, and provide cover. In providing supporting fires along streets, tanks were expected to occupy covered positions or operate only in areas controlled by motorized rifle units. During movement, tanks would move behind infantry at a distance beyond the effective range of enemy antitank weapons, but close enough to support the infantry with grazing fire from machine guns. The same principle was to be used for calculating the follow-on distance for other armored vehicles. Additionally, metal nets and screens were mounted 25 to 30 centimeters away from the armor to create protection from Chechen antitank rounds.

---

[27]A proper explanation of the concept of employing assault detachments requires a separate article.

Lesson Five: Anticipate and Resolve Communications Problems

As we have seen, a lack of training was the biggest problem for Russian troops and staffs in planning and executing the urban combat mission. The most significant technical problem was establishing and maintaining communications. In 1997–1998, no issue received more attention on the pages of the Russian army's most prestigious journal, *Armeyskiy Sbornik*. Obviously, this problem greatly complicated the execution of missions. If you can't coordinate and control units, how can they bring firepower to bear effectively?

The breakdown in communications occurred at the platoon, company, and battalion levels. Some of the problems were clearly the fault of Russian planners, such as the decision during the battle for Grozny to transmit all messages in the clear. This misstep obviously allowed the Chechen force not only to monitor all transmissions and thus prepare for what was coming next, but also to insert false messages in Russian communications traffic. Later, the Russians used message scramblers.[28]

The chief factor in the communications breakdown, however, was simply the vertical obstacles posed by urban structures. High-rise buildings and towers impeded transmissions, especially those in the high to ultra high frequencies. Communication officers had to consider the nature of radio wave propagation and carefully select operating and alternate frequencies, and they had to consider the interference caused by power transmission lines, communications lines, and electric transportation contact systems.[29]

---

[28]Valeriy Osyanin, Yevgeniy Komarov, and Nikolay Chistyakov, "Portable Radios in Grozny," *Armeyskiy Sbornik*, No. 10 (1996), pp. 41–43.

[29]Kudashov and Malashenko, p. 30. Other recommendations to improve communications included: using radios with automatic frequency tuning together with devices for guaranteeing scrambling and masking speech; using HF radios of armored vehicles with a supplementary receiver; using an "architectural waveguide" and "signal amplification by obstacle" plan (bouncing signals off of buildings or retransmitting them at intersections or via airborne platforms); locating VHF/UHF radios at a distance of three to five times the height of reinforced concrete upper stories or iron roof structures; putting antennas near windows or doors of upper stories when a radio is in a building; remoting radios from basements by using 10–15 meters of coaxial cable mounted to local objects with brackets or feeding into an existing television

Many radio transmitter operators were killed in the initial battles, as Chechens focused on soldiers carrying radios or antennas. To solve this problem, Russian radio operators began concealing their antennas. However, this led them to hide their whip antennas in a pocket or under a shirt and, in their haste to reassemble the radio while under fire, forget to reconnect the antenna.[30]

After-action recommendations by Russian communication specialists included developing more convenient and lighter-weight gear for radio operators, including wire-type antennas; outfitting units with cellular and trunk-adaptable radios; putting an indicator lamp on the radio sets to highlight problems; developing a common radio storage battery; and providing alternate antennas in follow-on models, capable of automatic connections in case primary antennas become disabled.[31]

The Russians noted that the Chechen forces used Motorola and Nokia cellular radios, and leased satellite channels on foreign relays. This enabled them to establish communications between base stations and to maintain quality mobile radio communications.[32]

Looking to the Future

When considering the initial failure of the Russian forces in Grozny in January 1995, it seems apparent that the issues outlined above, drawn from one-on-one discussions with participants as reported in Russian military literature from January to March 1995, would be at the center of the lessons-learned discussion. But Russian conclu-

---

antenna; and using beam antennas to maintain communications with a distant subscriber and a whip antenna for a radio net. Ibid.

[30]Osyanin et al., p. 41.

[31]Ibid., p. 43.

[32]Vladimir Komashinskiy, Valeriy Kumosov, and Nikolay Burenin, "In the Interest of Continuous Command and Control," *Armeyskiy Sbornik*, No. 10 (October 1996), pp. 48–49. Two other interesting articles on future communications techniques or adaptations are Aleksandr Anatolyevich Ivanov, "A View of the Future," *Armeyskiy Sbornik*, No. 11 (November 1996), pp. 6–8; and Nikolay Kochetov, Vladimir Artamonov, and Yevgeniy Komarov, "Flexible Command and Control," *Armeyskiy Sbornik*, No. 9 (September 1995), pp. 36–38. One report stated that some commercial cellular communications were used to support Russian formations and units in Grozny, but no further information was provided.

sions about the types of weapons and methods for attacking in future urban combat were much more imaginative than originally expected, focusing largely on the low end of the technological hierarchy. Such Chechen strategies as "defenseless defense" and "successive cities" seemed to force an innovative response.

For example, one of the lessons learned by Russian forces and underscored in their critique of combat in Grozny is the increasing utility of nonlethal weapons in future urban combat. This conclusion primarily refers to chemical weapons not banned by the Geneva Convention, such as tear gas and other agents.[33] The principal lesson Russian commanders seem to have learned is, "Don't fight this type of battle unless there is no other option." Gas is an option because it debilitates opponents and allows friendly forces to disarm them without lethal combat.

But the Russians are also considering high-tech, debilitating nonlethal means (rays of light causing blindness or seizures, subsonic sounds that penetrate concrete or metal and induce vomiting or spasms, electromagnetic waves, etc.).[34] New types of psychological operations, an old nonlethal technique, were under discussion as well. Psychological warfare techniques have been seriously upgraded through information technology developments. For U.S. forces concentrating on the high-technology solutions and approaches (firing around corners, devices to measure heartbeats through walls, etc.), it is important not to overlook counters to Russian innovations. At the very least, improvements should be made to U.S. soldier protective devices. Despite the standard U.S. aversion to the use of most of these Russian nonlethals, they may be more than attractive to other armies, especially if they are underbudgeted, undermanned, and undertrained.

The Russians learned other lessons from their Chechnya experience as well. They will likely now do everything in their power to persuade political figures to solve conflicts by peaceful means,[35] and their

---

[33]V. V. Kilunin, "New Views on Combat Actions in a City," *Military Thought*, Vol. 7, No. 3 (1998), pp. 41–43.

[34]Novichov et al., p. 164.

[35]Kilunin, p. 41.

preparation phase for urban combat probably will be comprehensive and exhausting in the future, since it is clear that Chechnya was not like Czechoslovakia. There will be more instruction on urban combat in their academies. The correlation of "other forces" (customs, religion, belief in the cause, receptivity to friendly forces, etc.) will be considered during the preparation phase, as will such factors as types of forces (guerrillas, regular, mercenaries), building materials, communications potential, local customs and resistance, friendly forces available, and the use of chemicals. Those chemicals may include "traction interrupters" to interfere with the working parts of equipment or to change a road surface, pyrophoric materials to burn nonflammables, or even biological materials to destroy electric and insulating materials.[36] It is clear that the Russian armed forces learned that if they can disable a person or piece of equipment, then it will be much easier to achieve their objective.

If force is used, there will be no preparatory fires (because it turns the population against you), but only supporting fires during the operation.[37] This concept may result in an extended use of direct fire artillery and a greater reliance on flame-throwers. It will be imperative to get civilians out of a city before fighting starts. Army aviation will be used to adjust artillery fires, provide battlefield command and control of troops, mark and coordinate boundaries, evacuate the wounded, and insert air assault forces at critical points in the city.[38] Finally, as the United States learned in Somalia, it is not always the best-equipped force that wins. Patience and discipline will play a greater role in the long run than the Russians acknowledged going into Grozny. This lesson must be learned by those who rely too heavily on precision weapons and think that victory is possible in the short term. Long-term engagement works against the intruding force; as civilian casualties mount, every move is scrutinized in the media, and the international community bands together to scold the "perpetrator."

Finally, a lesson learned by medical personnel and participants was the psychological stress of urban combat. Like the war in

---

[36]Novichov et al., p. 164.

[37]Kilunin, p. 41.

[38]Ibid., pp. 42–43.

Afghanistan, the Chechen conflict produced severe cases of combat stress and psychological trauma. A psychologically well prepared and trained Russian force was not available during the initial fight for Grozny. As a result, Russian commanders began establishing a reserve force only a few hundred meters away from the main force during the fighting. This reserve acted as a relief force that replaced the main force when it became psychologically spent. This usually occurred after about three hours of house and booby-trap clearing, which were the most stress-inducing activities other than clearing obstacles during the most intense days of the fight. A recent article about Chechnya noted that younger members of the native population there are also having serious troubles with stress-induced injuries from the war. One physician in Grozny, speaking about the children, noted: "They have become more aggressive, nervous, cruel. They have no respect for elders. They're dangerous to be around. They have psychological illnesses, terrible illnesses. Some can solve problems only with a gun."[39]

The first visible indicator of the traumatic nature of the attack on the Russian psyche was an article in *The Journal of Military Medicine,* just four months after the start of serious fighting. Major General V. S. Novikov, a professor in the medical service, gave a scathing account of the neurological disorders he was observing in Chechnya.[40] Novikov screened 1,312 troops in his survey. He found that 28 percent were healthy, while the other 72 percent had some type of psychological disorder (46 percent exhibited asthenic depression symptoms—insomnia, lack of motivation, anxiety, neuro-emotional stress, or tiredness—and the other 26 percent exhibited psychotic reactions such as high anxiety or aggressiveness, a deterioration of moral values or interpersonal relations, and excitement or depression). The longer a soldier was stationed in the war zone, as expected, the more radical the change in his neuropsychological condition. Novikov termed this condition Post-Traumatic Stress Syndrome, using the English acronym for this affliction in the Russian

---

[39]Edward Kline, "Chechnya after the War," *Central Asia Monitor-On-Line Supplement,* January 7, 1999.

[40]V. S. Novikov, "Psycho-physiological Support of Combat Activities of Military Personnel," *Military Medical Journal,* No. 4 (1996), pp. 37–40. The discussion and information in this section were taken from this article.

original. He had obviously studied the U.S. experience in Vietnam. The percentage of troops with combat-induced deficiencies was higher than in Afghanistan.[41]

Novikov's research also revealed that some 32 percent had experienced extreme stress while preparing for combat actions. These soldiers were taught active and passive muscular relaxation; others received psychological therapy or even pharmacological treatment (to treat insomnia or stress). After their removal from combat, troops' asthenic symptoms decreased while their psychotic disorders increased.[42]

Conclusions

The lessons of the fight for Grozny are several and sobering for anyone who contemplates using troops in an urban environment. While some of the lessons learned by Russian and Chechen combatants are peculiar to that region, others have wider applicability. No army wants to engage in urban combat, but increasing urbanization and the danger of strikes from high-precision weapons may well force the fight into the city, where the defender has all the advantages.

Preparation for urban combat should begin in peacetime. There is a vast array of possible courses of action, options, constraints, limita-

---

[41]Ibid., pp. 37–38.

[42]Ibid., pp. 38–39. Novikov recommended collective suggestive influences and pharmacological treatments as the most effective, and that there should be five specialists at army level (two psycho-physiologists, and one psycho-pharmacologist, one psychiatrist, and one medical psychologist). This group can assess 200–250 people per day and provide help. Ibid., p. 39. There were also urban combat "lessons learned" offered by some of the more savvy reporters who were present during the fight for Grozny that should be studied. Their experiences and opinions were colored differently than those of the combatants and remain valuable. Anatol Lieven, for example, offered three telling observations: that the effectiveness of even the best technologies for urban warfare will depend on how confused and afraid the man using them is; that Russians missed the capacity of Chechen social tradition to mobilize fighters and impose a discipline on them that goes beyond the "surface discipline" (imposed by basic training) of a modern army that does not provide nearly as strong a cause for the individual soldier to fight for (although it did not immunize young Chechens from PTSD); and that the Russian failure again demonstrates the limitations of firepower when fighting a dispersed infantry opponent behind good cover. Anatol Lieven, "The World Turned Upside Down," *Armed Forces Journal,* August 1998, pp. 40–43. Lieven's article is worth the time to look up and read.

tions, force mixes, enemy compositions, legal factors, and city characteristics that must be studied and understood. The most important point may be that there is no "standard urban combat operation." Each is unique to the opponent, the city, specific operational and tactical issues, and geopolitical considerations, among other factors. Understanding the elements and ramifications of urban combat is a difficult but crucial task for any army, but especially for one moving from a forward-deployed to an expeditionary state. In the latter case, the tasks required to sufficiently sustain or support urban combat are enormous.[43]

---

[43]The author would like to offer a special thanks to Mr. Les Grau of the Foreign Military Studies Office, Fort Leavenworth, Kansas, for his help with this manuscript.

ANNEX 4A: GROZNY II: LOGISTICAL SUPPORT DURING URBAN OPERATIONS
Mr. Lester W. Grau, FMSO

Russian Logistics and the Battle for Grozny

Lester W. Grau

Foreign Military Studies Office

(913) 684-5954

graul@leav-emh1.army.mil

Establishing Logistics Support

- Built on infrastructure of North Caucasus Military District
- Majority of logistics support facilities and units located near Mozdok garrison/railhead /airfield
- Built tent city of 3,000 tents with heaters, 114 mess halls, a shower and laundry train, shower and bath units, and vehicle wash points

- Established depots, supply dumps, and supply points from Mozdok to Grozny

- Established vehicle refueling points and rest stops with mess tents and heating tents on LOCs. Over 200,000 tons of POL used in first phase of conflict

- Established ad hoc traffic control service

Food Service

- Food provided 150% of normal daily ration per soldier—over 5000 calories
  - 300 grams meat
  - 50 grams heavy cream
  - 30 grams cheese
- Field bakeries-Mozdok, Vladikavkaz, & Kizlyar
- Three field bakeries north Grozny airfield— daily capacity of 8 tons of bread

- The KP-125 and KP-130 mess trailers could not be hauled behind ZIL-130 and GAZ-66 trucks in many areas during the thaw

- Mess trailers were hauled behind fuel and water trucks in combat formations

- Fuel trucks could not enter Grozny due to danger from small arms fire and shrapnel. Therefore, hot food had to be delivered in mermite containers carried in armored vehicles

- Troops often ate dry rations

High Demand Items

- Hand grenades
- Smoke grenades
- Demolition charges
- Grenade launchers
- Grappling hooks and ropes
- Light-weight ladders
- Night-vision equipment

- Smoke pots
- Smoke artillery rounds
- White phosphorus artillery rounds
- ZSU 23-4 and 2S6 ammunition
- Mortar ammunition
- Tear gas grenades
- Water

Transportation Support

- Primary long haul on train and plane.

- Prep period (11-30 Dec 94): Ground forces alone required 2850 long-haul trucks. Of these, 90 had serious breakdowns, 83 were written off.

- Battle for Grozny: Ground forces alone required 6700 long-haul trucks.

Railroad Troops

- Majority of heavy-lift into theater on rail

- Railroad troops restored 260 kilometers of track, cleared mines from 70 kilometers of main line, and repaired LOCs by restoring switches and electric power to the lines.

- Trains brought under mortar, artillery, and sniper fire

Supply Lessons Learned

- Winter weather necessitated additional specialized clothing and boots

- The majority of supply vehicles were not durable enough

- Fuel trucks could not enter Grozny due to danger from small arms fire and shrapnel. Fueling armored vehicles was a major problem

- Water supply was a major problem. Inadequate and purification methods were outmoded

- No place for crews to stow tents, sleeping bags, kits bags, field ranges, and heaters on the outside of combat vehicles

- Chronic need for an armored supply vehicle

- Needed at least a traffic control brigade during the preparation phase

- Captured POL stocks invaluable

Armored Vehicle Loss & Breakdown Rate

- Forward support maintenance repaired 217 armored vehicles

- Depot maintenance repaired 404 armored vehicles

- 225 armored vehicles written off

- 846/2221 armored vehicles (38%) out of action for some time during two-month fight

Repaired Vehicles: January 1995

- Two vehicle collection points established, one each axis-forward and rear

- 1286 vehicles repaired and returned

  - 404 armored vehicles

  - 789 wheeled vehicle

  - 75 artillery systems

The "Hangar Queen" Syndrome

- Prior to operation, 646 vehicles arrived needing repair

  - 41 artillery systems

  - 217 armored vehicles

  - 338 wheeled vehicles

Spare Parts and Accessories

- 573 tons for armored vehicle repair

- 605 tons for wheeled vehicle repair

- 60 tons for artillery system repair

Maintenance Lessons Learned

- Stop production of T-80 with gas-turbine engine

- Quality control of vehicles coming into theater required (prevent arrival of hangar queens)

- Large quantity of replacement vehicle accessories and spare parts needed

Medical Support

- Four special emergency medical treatment detachments dispatched in addition to TO&E medical support

- Companies reinforced with physician's assistant

- Battalions reinforced with doctor and ambulance section

- Regimental aid post where primary surgical treatment rendered

- Medevac helicopters to division hospital and beyond

Major Diseases

- Viral hepatitis
- Cholera
- Shigellosis
- Enterocolitis
- Diptheria
- Malignant anthrax
- Plague

Medical Lessons Learned

- Burns, shrapnel wounds, and sniper wounds are far more common in city fighting. Normal ratio of 3:1 or 4:1 wounded to killed is reversed to 1:3

- Armored ambulance needed for frontline service. BTR-80 usually pressed into service. Often had to wait until nightfall to evacuate the wounded due to snipers

- Forward air evacuation rare after several medevac helicopters shot down

- Conventional war treatment system much more effective in city fighting than counter-insurgency treatment system developed in Afghanistan

- Medical units forward in the city need to be protected and dug in. The best situation is the entire hospital in basements connected by trenches

- Clean water forward in quantity is essential

- 205th Separate Motorized brigade rendered combat ineffective by viral hepatitis (15% of force)

- Wounded usually seen by doctor in 15-20 minutes and reached surgical assistance within two-three hours

- Mortar was biggest casualty producer in city fighting

- Lice and skin diseases were a problem due to the difficulty in bathing and getting clean clothes

- Psychiatric casualties are high in urban combat. Units need to be rotated frequently.

Overall Logistics Lessons Learned

- No empty cities. Requirement to provide food, shelter, water, and medical treatment to civilians may outstrip same requirements for own force

- Requirement to restore sewage, water, electricity, and public order falls on military

- Rail and air transport essential

- All logistics functions are insufficiently staffed and supplied for urban combat

ANNEX 4B: SOFT LOG AND CONCRETE CANYONS: RUSSIAN URBAN COMBAT LOGISTICS IN GROZNY
Mr. Lester W. Grau and Mr. Timothy Lee Thomas, FMSO[44]

> Logistics make up as much as nine tenths of the business of war, and . . . the mathematical problems involved in calculating the movements and supply of armies are . . . not unworthy of a Leibnitz or a Newton.

—Martin Van Creveld[45]

Although logistics is a major concern of warfare, comparatively little has been written about it when compared to writings about the tactical and strategic aspects of various wars. As a subset, very little has been written about logistical support of urban combat. One historic precept of urban combat logistics is that ammunition expenditure increases dramatically when fighting in cities. Recent Russian experience in fighting for the Chechen capital city of Grozny in January/ February 1995 demonstrated that ammunition resupply was not the only problem. Demands on maintenance, supply, transport and medical support surpassed the capabilities of TO&E logistics units. Logistics demands were further increased by the requirement to provide humanitarian relief during the course of the fighting.

Russian tactics, techniques, and operational concepts for urban combat were based on their broad experience in the Great Patriotic War [World War II]. There were three underlying assumptions that shaped the Soviet/Russian concept of future urban combat: First, urban combat would be fought in nearly "empty" foreign cities where the bulk of the local civilian populace had left. Second, the enemy force in the city would be a conventional military force. Third, the army would have a period of conventional combat to fully develop procedures and identify problems before it began that most-

---

[44]This material will appear as Lester W. Grau and Timothy L. Thomas, "'Soft Log' and Concrete Canyons: Russian Urban Combat Logistics in Grozny," *Marine Corps Gazette*, Vol. 83 (October 1999), pp. 67–75, and it is printed with the kind permission of the editors.

[45]Martin Van Creveld, *Supplying War: Logistics from Wallenstein to Patton*, Cambridge: Cambridge University Press, 1986, p. 231.

difficult mission—fighting in a city. None of these assumptions proved correct in the fighting in Grozny. The civilians had no place to go and did not expect such extreme fighting, so they sat tight while the fighting engulfed the city. The Russian Army, as the sole government representative, was expected to provide food, shelter, clean water, sewage, electricity, and medical treatment to the civilians (who were citizens of the Russian Federation). The Russian TO&E combat service support units were barely able to sustain the Russian Army, let alone the large civilian populace, due to the increased demands of urban combat. It was beyond their capability and the civilians suffered. Eventually, the Russian Ministry of Emergency Situations (EMERCOM) helped restore these facilities.

Russian urban warfighting concepts were designed for fighting against another conventional army. The Chechen opposition were primarily guerrillas and irregulars backed by a small, fledgling regular force. The Chechens conducted a mobile "occasional" defense. They would hold one strong point one day and another on the next. The only exception was the Presidential Palace in the middle of the city, which they defended continually. Therefore, the Russian Army would stockpile supplies and munitions for projected attacks, but the attacks would frequently fall on empty buildings (or worse, buildings full of civilians). The enemy had moved, sometimes to the rear of the advancing Russian Army. This made it very difficult to direct combat service support to the critical sector in time. The Russian Army wanted to fight a linear battle, but the Chechen opposition made them fight a nonlinear battle. The Russian logistics units were unprepared for this.

The Russian Army began the fight in the capital city of Grozny—a modern city of 490,000 people mostly living in concrete and brick high-rise apartment buildings (an area over 100 square miles). The city is served by a major rail line, airfield, and is on a major highway net. Intercity movement relied on buses, trams, and private automobiles. Large factories and chemical plants competed with the oil industry for labor. A major oil and gas pipeline passes through the city. It is a difficult place for any army to begin a campaign.

Establishing the Theater Logistics Structure

The conscript-based Russian Army that entered the breakaway Republic of Chechnya in December 1994 was not prepared for the fight. There was not a single combat-ready division in the entire Russian force structure. Their deployed force was a composite grouping of various units that were rapidly cobbled together. The logistics units were in equally bad shape and were hurriedly assembled for the effort.[46] To further complicate logistics support, the ground campaign against the city of Grozny was mounted on three separate axes—from the west, northwest and east (map).[47] The fight for the city lasted one month, and cleanup operations took another month. The city was left in ruins.

Still, the Russian military planners and transportation personnel did an excellent job in assembling the composite force from all over Russia. Almost all the force and supplies initially traveled on rail or aircraft. Since Chechnya is part of Russia, the logistics buildup was founded on the existing logistics infrastructure of the North Caucasus Military District. The majority of the logistics support facilities and units were positioned near the Mozdok garrison. Mozdok has a good railhead and airfield and is located some 110 kilometers from Grozny. The Russian rear services built a tent city with some 3,000 heated tents, 114 mess halls, shower and bath units, and vehicle wash points. The rear services also brought a shower and laundry

---

[46]The Russian logistics system was a direct descendant of the massive Soviet logistics system—a system which, to a large degree, depended on the immediate transfer of trucks and drivers and other material from the civilian economy to the military upon mobilization. Russian industry had privatized and civilian trucks and drivers were no longer available—yet new arrangements had not been enacted. The use of composite logistical units further disrupted the logistics system, which was based on the premise that the higher echelon supplies the next two lower echelons. The echelons were no longer clearly defined, and the staffs at each level had little or no experience working together. It was a recipe for logistics failure.

[47]For a discussion of the battle for Grozny, see Timothy L. Thomas, "The Caucasus Conflict and Russian Security: The Russian Armed Forces Confront Chechnya. Military-Political Aspects and Military Activities, 11–31 December 1994," *Journal of Slavic Military Studies*, Vol. 8, No. 2 (June 1995), pp. 233–256; "The Caucasus Conflict and Russian Security: The Russian Armed Forces Confront Chechnya. Military Activities 11–31 December 1994," *Journal of Slavic Military Studies*, Vol. 8, No. 2 (June 1995), pp. 257–290; and "The Caucasus Conflict and Russian Security: The Russian Armed Forces Confront Chechnya. The Battle for Grozny, 1–26 January 1995," *Journal of Slavic Military Studies*, Vol. 10, No. 1 (March 1997), pp. 50–108.

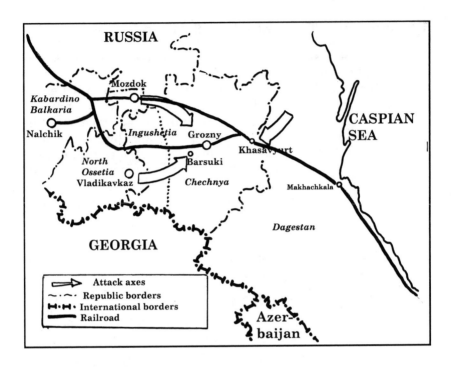

train forward to Mozdok.[48] Long-haul was by rail and air, and soon depots, supply dumps, and supply points were established at Mozdok extending toward Grozny. Three truck LOCs were established— one per main axis. Vehicle refueling points with rest stops containing mess tents and heating tents were set up along the LOCs.[49] Trucks were essential to move supplies from the airfield and railhead forward toward Grozny.

Food Service

The fighting for Grozny began on New Year's Eve 1994. Chechnya is mountainous and the winters are cold and snowy. The planners

---

[48]Mikhail Shchepakin, "Tyl v Chechne" [Rear Service in Chechnya], *Armeyskiy sbornik* [Army digest], June 1995, p. 20. Unfortunately, the front-line troops seldom were able to use the laundry and bath facilities. As a result, skin diseases and lice were a problem among combatants.

[49]Ibid.

decided to provide 150 percent of the normal ration to each soldier. This would exceed 5,000 calories and included a daily 300 grams [10.5 ounces] of meat, 50 grams [1.75 ounces] of heavy cream, and 30 grams [1.05 ounces] of cheese. Field bakeries were established on each of the main axes at Mozdok, Vladikavkaz, and Kizlyar. Later, when the north Grozny airfield was captured, the Russians positioned three field bakeries there—with a daily capacity of eighteen tons of bread.[50] There should have been plenty of food for every soldier.

However, the Russians had trouble delivering rations to the forward fighting positions. Meals were prepared on the KP-125 and KP-130 mess trailers. These are very serviceable cooking units that are hauled by the ZIL-130 or GAZ-66 trucks. However, when the ground around Grozny thawed, these trucks could not haul their mess trailers through the soupy mud. Then, the only way to haul the mess trailers forward was behind fuel or water trucks. Fuel trucks could not enter the city, since a single bullet might set the entire vehicle ablaze, so mess trailers often got no further than the outskirts of the city. Therefore, the food had to be ladled into mermite-type containers, which were then loaded into armored personnel carriers for transport into the city.[51] This absence of "hard log" transportation was a constant problem.

Often the troops at the forward positions had to eat dry rations.[52] These dry rations did not provide the minimum daily required amount of calories or vitamins.[53] Often, the troops who needed the extra calories the most were not even getting the minimum daily

---

[50]Shchepakin, pp. 20–21.

[51]Ibid., p. 21. A mermite container is a U.S. Army insulated food container.

[52]Dry rations are similar to the old U.S. Army C-ration. There are three types of dry rations. The first contained a can of meat, some crackers or toast, some jam, and a tea bag. The second contained two cans of meat mixed with oatmeal. The third contained a can of meat and a can of vegetables or fruit. Lester W. Grau, *The Bear Went Over the Mountain: Soviet Combat Tactics in Afghanistan,* London: Frank Cass Publishers, Inc., 1998, p. 5.

[53]Lester W. Grau and Michael A. Gress, *The Bear Looks Back: A Russian General Staff Retrospective on the War in Afghanistan,* manuscript awaiting publication.

requirement. Thus, the initial plan to provide 5,000 calories per day went widely astray, primarily due to inadequate transport.[54]

High-Demand Items

Besides small arms ammunition, front-line infantry used copious amounts of hand grenades, smoke grenades, smoke pots, demolition charges, flame thrower rounds, RPG-7 rounds, and single-shot disposable antitank grenade launchers. Tear gas grenades were often required at certain points on the battlefield and had to be pushed forward. The front-line infantry also had an immediate need for quantities of grappling hooks and ropes, lightweight ladders, and night-vision equipment. Many of these items were delivered by emergency airlift to Mozdok. Where there was a shortage of night-vision equipment, the Russians used mounted and dismounted searchlights to illuminate the battlefield and dazzle the Chechen opposition.[55]

Mortars produced the most casualties on both sides, and HE and smoke mortar ammunition was always in demand. Artillery was also used, often in a direct fire role. One-fifth of the artillery ammunition fired was smoke or white phosphorous—consequently, these were high-demand items. Smoke screened infantry movement, and white phosphorus smoke had the additional advantage (or disadvantage) of being lethal, capable of penetrating existing protective mask filters, and not banned by any international conventions.[56]

One of the most effective Russian weapons in city fighting was the venerable ZSU 23-4—a lightly armored self-propelled antiaircraft gun whose four 23mm barrels spat out up to 3,200 rounds per minute. The elevation and deflection of the system, as with its modern equivalent, the 2S6, provided an excellent countergunner

---

[54]There have also been accusations that much of this food was diverted to the Russian black market.

[55]Sergey Leonenko, "Ovladenie gorodom" [Capturing a city], *Armeyskiy sbornik* [Army digest], March 1995, p. 32.

[56]Ibid., p. 33.

weapon for city fighting. However, keeping the ZSU 23-4 and the 2S6 in 23mm and 30mm ammunition was a constant problem.[57]

Clean drinking water was a high-demand item, but delivery of clean water forward often proved too difficult. Individual water treatment panticides took too long to work. Fighting is thirsty work, and soldiers drank what was available. Viral hepatitis and cholera were the result.[58]

POL was critical, as the Russians used over 200,000 tons of POL during the battle for Grozny. Captured POL stocks proved very useful to the Russian ground forces. A major problem was moving the POL stocks up close to the units in contact.[59]

Transportation Support

Primary heavy-lift long-haul into the theater was on rail. Railroad troops had to restore 260 kilometers of track, clear mines from another 70 kilometers, repair switches, and restore electric power to electric rail lines. Trains had to be protected as they came under mortar, artillery, and sniper fire.[60]

Air transport played a significant role in the long haul of men and supplies. High-demand items were almost always shipped by air. Practically the entire Russian Military Transport Aviation (VTA), plus some commercial aviation, was involved in supporting the effort.

In theater, truck transport was essential. During the short preparation period (11–30 December 1994), 2,850 long-haul trucks sup-

---

[57]Vladimir Suzdal'tsev, "Chechenskie uroki voyskovoy PVO" [Air defense lessons from Chechnya], *Armeyskiy sbornik* [Army digest], September 1995, pp. 23–24.

[58]P. I. Ogarkov, V. V. Malyshev, S. A. Tsutsier, and N. V. Mikhaylov, "Epidemiologicheskaya kharakteristika i laboratomaya diagnostika virusnykh gepatitov v federal'nykh voyskakh na territorii chechenskoy respubliki" [Epidemiologic characteristics and laboratory diagnosis of viral hepatitis among federal forces deployed in the Chechen Republic], *Voenno-meditsinskiy zhurnal'* [Military-medical journal], August 1996, p. 48.

[59]Pavel Gorupay, "Sluzhba goryuchego v Chechenskom krizise" [POL support in the Chechen crisis], *Armeyskiy sbornik* [Army digest], May 1995, p. 37.

[60]Vadim Fedotov, "Magistral" [Magistral], *Armeyskiy sbornik* [Army digest], March 1995, pp. 48–49.

ported the ground forces. Of these, 90 had serious breakdowns and 83 were written off as not economically repairable. During the battle for Grozny, the long-haul truck requirement for ground forces increased to 6,700 trucks.[61] Controlling all this traffic was a problem. The Russians had forgotten about their Afghanistan experience, where the Soviet 40th Army had a traffic-control brigade assigned to control convoys. Consequently, the Russians had to assemble an ad hoc traffic-control brigade at the same time that they were conducting a major operation.[62] Traffic control is just one example of where the lack of adequate time in the preparation phase can cripple the entire effort. Getting logistics in place and ready takes time. Political leaders did not give that time to the Russian Army.

The fighting in Grozny highlighted several problems. Supply trucks were soft-skinned, not rugged enough, and could not be exposed to urban combat. One of the major problems with supplying forward forces was that trucks could only go forward to a certain point. Then all the cargo had to be transloaded onto BTRs, MTLBs, or other armored vehicles. The armored vehicles were not designed primarily for carrying cargo and had to make several trips to haul a single truck's load. This meant that the combat commander lost the use of many, if not most, of his armored combat vehicles for combat. They were busy hauling ammunition, food, and water or serving as ambulances. There was a chronic need for an armored supply vehicle that could move right up to the forces in contact.[63]

Connected with this problem was the lack of load-carrying racks on the outside of Russian armored vehicles. Tents, sleeping bags, kit bags, squad stoves, and the like were carried in the supply trucks. The trucks could not get forward and there was no place to carry soldiers' gear on or in the armored vehicles. As a result, combatants had to do without individual gear for days at a time.

---

[61]Nikolay Kovalev, "Avtotekhnika v chechenskom konflikte" [Wheeled vehicles in the Chechen conflict], *Armeyskiy sbornik* [Army digest], March 1996, p. 62.

[62]Valeriy Buravtsev, "Na voennykh dorogakh" [On military roads], *Armeyskiy sbornik* [Army digest], April 1995, p. 40.

[63]In the late 1980s, the Soviets designed a prototype armored supply truck, but it apparently was not produced in quantity before the collapse of the Soviet Union.

Rearming and refueling combat vehicles was particularly difficult. It usually had to be done at night. Rearming and refueling on site meant that lots of soldiers carried fuel cans and ammunition boxes forward—a long, arduous, and hazardous process. Withdrawing vehicles, particularly tanks, to rearm and refuel is also difficult.[64] Forward-deployed troops did not always get the word that their supporting armor was being withdrawn only for rearming and refueling, sometimes misinterpreting a withdrawal of tanks as part of a general withdrawal.

Maintenance

Maintenance requirements exceeded expected maintenance norms for conventional combat during the two-month urban fight.[65] Armored vehicle maintenance was especially critical, and unit maintenance officers tried to keep control of their vehicles and repair as many vehicles as possible at regiment or brigade level. Still, during the two-month fight, forward support maintenance repaired some 217 armored vehicles, depot maintenance repaired some 404 armored vehicles, and 225 vehicles were written off as being non-repairable. Thus some 846 of 2,221 armored vehicles involved in the fight (38 percent) were out of action for some period of time—although not simultaneously.[66] Combined with the armored vehicles detailed for supply runs and medical evacuations, some combat commanders were lucky to have 40 percent of their armored vehicles present for combat.

---

[64]A good example from another war is Ali Ahmad Jalali and Lester W. Grau, *The Other Side of the Mountain: Mujahideen Tactics in the Soviet-Afghan War*, Quantico: U.S. Marine Corps Study, 1998, pp. 198–204.

[65]The authors are unable to find any indication that the Russians had meaningful, contemporary logistics norms for urban combat. Their great urban battles of World War II (Stalingrad, Warsaw, Konigsberg, Budapest, Berlin) were fought primarily with foot infantry and little mechanized support. The mechanized forces were used to encircle the cities and fight on the city approaches.

[66]Sergey Maev, "STO v Grozny" [Technical Maintenance Stations in Grozny], *Armeyskiy sbornik* [Army digest], December 1995, p. 58. For a look at how the 225 armored vehicles were destroyed, see Lester W. Grau, "Russian-Manufactured Armored Vehicle Vulnerability in Urban Combat: The Chechnya Experience," *Red Thrust Star*, January 1997, pp. 16–19.

In order to meet increased maintenance demands, the Russians formed three separate maintenance battalions and two maintenance detachments in addition to the deployed TO&E units.[67] The Russians established collection and repair points on each axis. In the west, the rear point was in Vladikavkaz, while the forward was located near the trains of an airborne division. In the north, the rear point was in Mozdok, while the forward was with the trains of a motorized rifle brigade. In the east, the Russians established three forward collection and repair points: with the trains of a motorized rifle division, an airborne regiment, and a motorized rifle regiment.[68] During the month of January 1995, forward support and depot maintenance repaired 1,286 vehicles and returned them to their units. These included 404 armored vehicles, 789 wheeled vehicles, and 75 artillery pieces. Maintenance personnel evacuated another 259 damaged armored vehicles from Grozny during January fighting. Due to the complexity of fire control systems, automatic reloading systems, electric systems, and communications systems, 26 percent of some types of armored vehicles had to be repaired by factory representatives.[69]

Combat damage and equipment failure were not the only maintenance problems. Money was not available to repair many vehicles prior to the war, and so 646 "hangar queens" were shipped into the theater. All these 646 vehicles (338 wheeled vehicles, 217 armored vehicles, and 41 artillery pieces) had to be repaired prior to the initiation of combat. Maintenance demands exceeded norms to such a degree that 573 tons of armored vehicle spare parts and accessories, 605 tons of wheeled vehicle spare parts and accessories, and 60 tons of artillery spare parts and accessories had to be brought into theater to supplement the on-hand repair parts.[70] As a result of its poor performance and high fuel consumption during the fighting in

---

[67]Maev, p. 55.

[68]Ibid.

[69]Ibid., p. 58. Factory representatives were especially needed to repair the T-72 and T-80 tanks; the 2S1 SP 122mm and 2S19 152mm SP howitzers; the 2S5 SP 152mm gun and the 2S6 air defense system.

[70]Ibid.

Grozny, the Russian high command canceled production of the gas-turbine engine for the T-80 tank.[71]

Medical Support

Russian Army care of the wounded was usually well planned and executed once the patient reached the battalion aid station. Three weeks before the Russian incursion, the Russian Army established and trained special emergency medical treatment detachments in each military district. Four of these detachments deployed to Chechnya to support the maneuver units and supplement their TO&E medical units.[72]

The Russians utilized their normal conventional war evacuation system and usually employed ground medical evacuation as the quickest and safest form of evacuation. Each maneuver company was reinforced with a physician's assistant, and each maneuver battalion had a medical doctor plus the ambulance section. Surgeons, anesthetists, and additional nurses manned the regimental medical post.[73] Wounded were normally evacuated to the regimental medical post by makeshift armored ambulances (BTR-80), since the Chechens fired on the soft-sided ambulances. Forward medical stations and hospitals needed to be dug in or deployed in basements, as the Chechens also shelled these. Patients requiring more extensive medical care were evacuated by MEDEVAC helicopter and MEDEVAC aircraft.[74] Forward air evacuation was not used much, particularly after the Chechens shot down several MEDEVAC helicopters. The fighting in Grozny proved the need for a specially designed armored ambulance.[75]

---

[71]Mikhail Zakharchuk, "Uroki Chechenskogo krizisa" [Lessons of the Chechen crisis], *Armeyskiy sbornik* [Army digest], April 1995, p. 46.

[72]N. N. Novichkov, V. Ya. Snegovskiy, A. G. Sokolov, and V. Yu. Shvarev, *Rossiyskie vooruzhennye sily b chechenskoim konflikte: analiz, itogi, vyvody* [Russian armed forces in the Chechen conflict: analysis, results and outcomes], Moscow: Holweg-Infoglobe-Trivola, 1995, p. 131.

[73]Ibid., p. 132.

[74]Ibid., p. 134.

[75]Yuri Savvin, "Za zhizni voinov" [For the lives of the warriors], *Armeyskiy sbornik* [Army digest], March 1995, p. 45.

City fighting produced a different percentage of casualty types. Red Cross statistics for limited conflicts usually reflect 23 percent wounded from mines, 26 percent from bullets, 46 percent from shrapnel, 2 percent from burns, and 3 percent miscellaneous. In the city fighting of Grozny, however, there was a higher percentage of burns, and the majority of wounds were caused by mortar fire. The majority of those who were killed or died from wounds were hit in the head and chest by sniper fire (particularly among the civilians who did not have flak jackets and helmets). Whereas the normal ratio of wounded to killed is 3:1 or 4:1, this was reversed in the Grozny city fighting, where three were killed for every wounded. [This ratio is probably skewed and reflects that many of the wounded could not be reached and given first aid in time. The actual initial ratio was probably closer to 2:1 wounded to killed.] Snipers presented a problem for medical evacuation, and frequently the wounded could not be evacuated until nightfall.[76]

The Russian Army record in disease prevention was nowhere near as impressive as its handling of the wounded. Russian soldiers frequently lacked clean drinking water, clean clothing, hot rations, and washing facilities. Personnel suffered from viral hepatitis, cholera, shigellosis, enterocolitis, diphtheria, malignant anthrax, and plague. One combat brigade had 240 simultaneous cases of viral hepatitis. Since Russian field units were down to 60 percent strength or less at this time, a brigade would be lucky to muster 1,500 personnel. Over 15 percent of this one brigade was down with hepatitis. The brigade was combat ineffective due to disease, and contaminated water was the main culprit. Bacilli from the human intestinal tract were present in 60 to 80 percent of dishwater tested. Some 4 percent of the sick worked in food handling or water distribution.[77]

Psychiatric casualties are higher in urban combat. Most of the fighting in Chechnya was in cities (first in Grozny and then a succession of smaller cities and finally towns). A Russian military psychia-

---

[76]Novichkov, p. 133. For further reading, see Lester W. Grau and William A. Jorgensen, "Handling the Wounded in a Counter-Guerrilla War: The Soviet/Russian Experience in Afghanistan and Chechnya," *U.S. Army Medical Department Journal,* January/February 1998, pp. 2–10.

[77]Lester W. Grau and William A. Jorgensen, "Viral Hepatitis and the Russian War in Chechnya," *U.S. Army Medical Department Journal,* May/June 1997, pp. 2–5.

trist conducted a survey of 1,312 soldiers during the combat.[78] Soldiers surveyed were still capable of performing combat functions. The survey found that 28 percent were healthy and the other 72 percent had some type of psychological disorder (46 percent exhibited depression; a weak, apathetic, or retarded motor state; or simple insomnia). Other disorders in the 46 percent included a lack of motivation, high anxiety, neuroemotional stress, tiredness, and hypochondriacal fixation or panic attacks. The other 26 percent exhibited psychotic reactions such as high anxiety or aggressiveness, and a deterioration of moral values or interpersonal relations, excitement, or acute depression. About 40 percent of the soldiers screened demonstrated a lack of neuropsychological stability. The longer a soldier was stationed in the war zone, the more radical the change in his neuropsychological condition. The percentage of troops with Post-Traumatic Stress Syndrome (PTSD) was higher than in Afghanistan—reflecting the impact of urban combat.[79]

The Russians noted that they should have rotated units frequently to allow the soldiers to bathe, sleep, train, and readjust. This would have required much larger reserves than were available and created an additional logistics load. The Russians recommended that future urban combat include more psychiatric support—including professionals who would work forward in the units.[80]

Prisoners and Detainees

Separating combatant from noncombatant was a difficult problem for the Russian armed forces. They began by simply examining suspects for bruises in the pocket of the shoulder to see if they had fired

---

[78]V.S. Novikov, "Psikhofiziologicheskoe obespechenie boevoy deyatel'nosti voennosluzhashchikh" [Psycho-physiological Support of Combat Activities of Military Personnel], *Voenno-meditsinskiy zhurnal* [Military Medical Journal], No. 4, April 1996, pp. 37–40.

[79]Ibid., pp. 37, 38. The unanswered question is, What were these soldiers like before combat? The Russian Army is a conscript army that draws from the segment of society not smart or connected enough to avoid conscription. An inordinate number of conscripts have dropped out from school, have criminal records, and are in poor health. In many units the Russian barracks environment is stressful, with severe hazing and physical abuse viewed as a normal part of army life. Many of the Russian soldiers were probably "damaged goods" before they arrived in Grozny.

[80]Ibid., p. 39.

a weapon and looking for powder or burn marks on suspects' fore-arms and shirt cuffs. By the second month, the Russian internal troops resorted to a simpler method: rounding up most Chechen males and putting them in "filtration" camps. The camps were de-signed to identify and separate those Chechens who had possibly fought against the Russians from peaceful civilians. Prisoner gather-ing and maintaining filtration camps, run by the Ministry of Internal Affairs (MVD), required a considerable amount of vehicles, food, POL, water and security support—much of which apparently came from Ministry of Defense assets. The Russians were not prepared to handle the mass of prisoners. As a consequence, the prisoner situa-tion was so disordered that the International Red Cross had difficulty locating camps and found it impossible to trace individual prison-ers.[81]

Handing Off Support to Government and Nongovernment Agencies

The Russian Armed Forces could not simultaneously fight and re-store food delivery, sewers, water processing, public health, and public services in the city. This task was eventually handled by the Ministry of Emergency Situations (EMERCOM). EMERCOM is the successor to the Soviet Union's civil defense organization. It is a rough equivalent to the U.S. Federal Emergency Management Agency (FEMA).

EMERCOM managed to do a great deal to restore vital services to Grozny. The EMERCOM technical directorate dispatched its epi-demic prevention service, which monitored/inspected food supplies; performed bacteriological testing on the sick; conducted disinfec-tion, disinfestation, and rat control in the city; educated the public on health issues; and restored Chechnya's epidemiological and health centers. It also set up water distribution points at three hospi-tals; equipped three bathing facilities for patients and medical per-sonnel; removed fallen debris and regular garbage; put three hospital cafeterias back in order; delivered medical and equipment and

---

[81]Erik Reumann, "Red Cross Ready to Return," *Passport*, July/August 1996, p. 54. There are allegations of vile treatment and summary execution of prisoners by both sides. Both sides clearly committed gross violations of human rights.

drugs; provided all hospitals with a fully equipped ambulance; restored a maternity center; and provided 190 oil heaters.

EMERCOM technical services further restored more than 50 kilometers of high-voltage power lines; restored three heat and power plants; and set up eight diesel electric stations and repaired another. It restored 5,331 meters of gas lines; delivered gas to 34 high-rise apartment buildings and 21 private buildings; and restored a gas distribution point in the center of Grozny. It set up a field bakery and delivered bread-making ingredients.

Finally, EMERCOM cleared mines from five water pumping and purification stations; performed engineer and medical reconnaissance of water sources; set up water-collection stations on the Sunzha River; provided emergency repair to water and sewage systems, and restored 21 damaged segments on major waterlines on eleven streets. EMERCOM also inspected and tested Grozny's radioactive and chemical sites; tested fifteen dangerous interagency sites; and exchanged NBC information with the Interior Ministry.[82]

Beginning in January 1995, EMERCOM assisted UN agency assessment efforts to help displaced persons. As a result, the UN issued a "Flash Appeal" for immediate assistance in February. The appeal requested $25.1 million for shelter, water, sanitation, food, health, community services, distressed children, and support.[83] Expected donors included the UN High Commission on Refugees (UNHCR), the UN International Children's Emergency Fund (UNICEF), the World Food Program (WFP), the World Health Organization (WHO), and the International Migration Organization (IOM). Although all the requested aid was not given, these organizations eventually aided some 220,000 people (200,000 from Chechnya, and the rest from North Ossetia plus some from Georgia). Further, the International Red Cross distributed some 250,000 food packages monthly, established a soup kitchen in Grozny, reopened a hospital in Grozny, and

---

[82]All material in this section were the result of conversations between Mr. Thomas and the Public Affairs Office of EMERCOM.

[83]"United Nations Consolidated Inter-Agency Appeal for Persons Displaced as a Result of the Emergency Situation in Chechnya," United Nations, March 1995, p. iii of the Executive Summary.

opened a "contact service" where people could apply to reestablish contact with lost relatives and friends.[84]

Conclusion

The Russian Army was poorly prepared for combat in Grozny. It muddled through and even improved somewhat over time. However, the lessons learned from its combat are not limited to the Russian Army, but apply to any modern, mechanized force fighting a determined enemy in a city. The logistics lessons also apply. Urban combat will demand increased amounts of ammunition and special equipment, yet a major problem will be getting the supplies forward to where they are needed. There is a need for a rugged, armored supply/medical evacuation vehicle and a better way of rearming and refueling combat vehicles in the forward area.

There are no empty cities, and the ground commander should conduct contingency planning in case he must care for the needs of the civilian population and restore critical services. The military commander may become the de facto city manager and should be prepared to keep the civilian populace alive and healthy, should this be required. To limit the time spent in this area, the commander should learn to work effectively with other government and nongovernment agencies. This means that a higher percentage of combat service support personnel may be needed before combat begins. Rail and air transport are critical to the logistics effort. Port and rail rehabilitation units may need to be among the first units into an urban theater.

Currently, logistics units in many armies are insufficiently staffed and equipped for urban combat. Urban combat greatly stresses ammunition, water, food, and POL resupply. Maintenance demands greatly increase during urban combat. Vehicle evacuation/obstacle clearance will be an essential engineer/maintenance task. Factory representatives will need to accompany the force, additional maintenance units are needed, and additional spare parts will need to be on hand prior to the initiation of combat. An aggressive screening program is required to keep "hangar queens" out of the theater.

---

[84]Reumann, p. 54.

Medical support will also require reinforcement, and preventive medicine will play a major role in preserving the force, as water-borne diseases are very dangerous to its well being. Mortar wounds, burns, and psychiatric trauma increase dramatically in urban combat.

None of these are exclusively Russian problems or observations. Actually, the Russian logistics services performed reasonably well considering the monumental handicaps they had to overcome. The extensive logistics system designed in the Soviet era was no longer in place, and the Russian Army did not have a viable replacement. Logisticians had just over a two-week buildup phase with no logistics rehearsals. [The Soviet Army conducted six major exercises to prepare for the invasion of Prague in 1968. Logistics rehearsals were an essential element of these exercises.][85] The political masters gave the commanders no time to develop the theater, although there was no military reason to hurry. The logistics units were often composite units cobbled together on site. There were few habitual relationships among the participating staffs. Equipment was often broken on arrival. The logisticians were supporting a battle for which the planning norms were outdated. They were faced with the challenge of caring for a large civilian populace while other government and non-government agencies, which could help deal with this problem, were slow to arrive. Yet Russian logisticians adapted to their shortcomings and provided adequate support under very trying circumstances. Other armies would be wise to study the difficulties they encountered and adjust accordingly.

---

[85]Graham Turbiville, "Soviet Bloc Maneuvers," *Military Review*, August 1978, pp. 19–35.

ANNEX 5: Q&A FROM RECENT OPERATIONS PANEL

Question: Is the culture of the Serbs similar enough to that of former Warsaw Pact nations that we're going to have the same attitude regarding the low value of life during operations in the cities of Kosovo?

Mr. Thomas: I personally call it the Son of Sam psychosis. It happened in Bosnia earlier. This killing begets killing. And it keeps going and going and going. We had a Russian down at a NATO PSYOP course talking about human factors; he said people are kind of like squirrels. When we get too many of them, we've just got to thin out the population. It's a very different cultural mindset. I'm certainly not sitting here telling you that all Russians are like this or that all Serbs are like this, but that comment from him and another one made right after Grozny reinforced the point that there is a diverse cultural difference in the way some people approach life and death. Over in Bosnia they were putting people on table saws; the horrific nature of the way they were killing just stunned us. There are sick folks out there and I think there's a lot of sick warriors in the world right now. I personally don't know how we're going to stop them.

Mr. Grau: When you're fighting for national survival, the survival of your culture, you're going to fight a lot harder because you're fighting the ultimate battle.

Question: Did Haitian cultural considerations impact your decision to use nonlethal weapons?

COL Patton: The situation down in Haiti was very bleak; it still is today. There's a lot of vigilante justice down there and they are a somewhat violent people. What we had going for us and one of the reasons that we chose to use nonlethal weapons was that they love Americans for the most part. They had a great affinity for our presence down there. We had a significant PSYOP campaign to let the country know that we were down there for benevolent reasons. We were repairing schools and hospitals and digging wells. We felt that the nonlethal weapons provided us with an option short of deadly force that would probably serve us well in most situations. As I mentioned, we never had to use them while I was there, but we were prepared to use them for crowd control because we did get gatherings of large groups around our compounds.

Question: What's the most urgent need for young soldiers and marines right now regarding deficiencies in equipment and organization given our discussions over the past two days?

COL Patton: As I sit here thinking about my experience in Haiti, in that particular environment I don't know that we really had a deficiency, especially of an equipment nature. I think our equipment was adequate. I think what that environment proved to us is that well-disciplined, well-trained, well-led troops can still accomplish missions in that type of environment. It was nothing like what's been described in previous discussions today about city fighting, urban combat. We just didn't face that type of threat. We had good soldiers and good marines down there.

Glenn, Russell W., et al., *Denying the Widow-maker: Summary of Proceedings, RAND-DBBL Conference on Military Operations on Urbanized Terrain*, Santa Monica, CA: RAND, 1998.

Glenn, Russell W., "Fox Trot: Seeking Preparedness for Military Urban Operations," *Armed Forces Journal International*, Vol. 136 (May 1999), pp. 46–49.

Grau, Lester W., and Timothy L. Thomas, "'Soft Log' and Concrete Canyons: Russian Urban Combat Logistics in Grozny," *Marine Corps Gazette*, Vol. 83 (October 1999), pp. 67–75.

Hahn, Robert F., and Bonnie Jezior, "Urban Warfare and the Urban Warfighter of 2025," *Parameters*, Vol. 29 (Summer 1999), pp. 74–86.

Thomas, Timothy L., "The Caucasus Conflict and Russian Security: the Russian Armed Forces Confront Chechnya III. The Battle for Grozny, 1–26 January 1995," *Journal of Slavic Military Studies*, Vol. 10 (March 1997), pp. 50–108.

Thomas, Timothy L. "The Battle of Grozny: Deadly Classroom for Urban Combat," *Parameters*, Vol. 29 (Summer 1999), pp. 87–102.